Sustaining Coastal Zone Systems

For details of other Dunedin Academic Press publications please see
www.dunedinacademicpress.co.uk

Sustaining Coastal Zone Systems

Edited by

Paul Tett, Audun Sandberg & Anne Mette

DUNEDIN

Published by
Dunedin Academic Press Ltd
Hudson House
8 Albany Street
Edinburgh EH1 3QB
Scotland

ISBN 978 1 906716 27 1

© 2011 Dunedin Academic Press

British Library Cataloguing in Publication Data
A catalogue record for this book is available from the British Library

Typeset by Makar Publishing Production
Printed and bound in Poland by Hussar Books

Contents

	List of figures	vii
	List of tables	ix
	Biographical information	xi
	Editors Preface	xvii
Chapter 1	Introduction *Paul Tett and Audun Sandberg*	1
Chapter 2	Coastal zone problems and policy context *Audun Sandberg*	29
Chapter 3	The Systems Approach *Paul Tett, Anne Mette, Audun Sandberg and Denis Bailly*	53
Chapter 4	Modelling coastal systems *Paul Tett, Maurizio Ribeira d'Alcalà and Marta Estrada*	79
Chapter 5	Bridging the gap between science and society *Anne Mette*	103
Chapter 6	Conclusions *Paul Tett, Anne Mette, Audun Sandberg, Marta Estrada, Maurizio Ribeira d'Alcalà, Tom Sawyer Hopkins and Dennis Bailly*	137
	Glossary and Acronyms	151
	References	157
	Index	167

List of figures

1.1   The shore at Scheveningen in the 17th century.   5

1.2   Humanity becoming distinct from natural ecosystems.   10

1.3   A social-ecological system.   11

1.4   Map of Europe showing the 18 sites at which coastal zone
      problems were studied during SPICOSA.   21

1.5   The steps of SPICOSA's Systems Approach Framework or SAF.   22

1.6   The Science–Policy Interface   23

2.1   Puffins on the Island of Hornøya, Norway.   30

2.2   Cartoon of ecosystem collapse.   46

3.1   The Lagoon of Venice, pictured from a satellite.   54

3.2   A coastal zone system showing boundaries and boundary conditions.   55

3.3   Simple input–output response curves.   58

3.4   Logic diagram for a Negative feedback loop.   60

3.5   Complex input–output response curves.   63

3.6   Panarchy: the adaptive cyclical character of an
      exploitation–release system.   64

3.7   Scenarios on a flipchart.   69

3.8   Results of the scenarios of the Millennium Ecosystem Assessment.   74

3.9. Augmented black box.   74

4.1   Picture of a clockwork orrery.   80

4.2   Simple conceptual model: phytoplankton, clams and clam fishery.   84

4.3. Conceptual Model for processes related to a single clam.   87

4.4   Conceptual Model for (mainly economic) processes related to
      the clam fishery.   87

4.5   More complex top-level conceptual model.   88

4.6   Using STELLA® to model algal growth processes.   93

4.7   Testing a model. A simulation of phytoplankton chlorophyll
      in the Scottish fjord, Loch Creran, for 1975.   96

4.8   Example of an ExtendSim® block.   97

4.9   Institutional map for aquaculture in the Lagoon of Venice.   99

5.1   Newspaper articles about people charged for illegal fishery
      and a public meeting which ended in a fight.   106

5.2   Diagram about the collaborative interfaces in a SAF to illustrate the
      communication flows and collaboration processes within a SAF.   114

5.3   Deliberation sheet for a deliberation matrix.   115

5.4   Elements of a Deliberation Matrix tool.   116

5.5   Part of an institutional map for the Issue of eutrophication
      in the Swedish Himmerfjärden.   120

5.6   Conceptual model of the freshwater governance in the
      Charente river catchment and its coastal zone.   131

5.7   Animated mapping of Charente river water flow level at
      monitoring stations for irrigation regulation.   132

6.1   The Coastal-SAF website that provides further information
      on the topics in this book.   147

Picture Credits

We are grateful to:

1.1   Wikipedia Commons, from a collection of public-domain reproductions
      by the Yorck Project, 2002; original painting in Staatliche Museum Kassel,
      Germany.

2.1   Wikipedia Commons, picture by Hanno in 1996, released under the GNU
      Free Documentation Licence and the cc-by-sa-2.5 Licence.

3.1   NASA/GSFC/MITI/ERSDAC/JAROS, and U.S./Japan ASTER Science Team.

4.1   Wikipedia commons, author: Kaptain Kobold in 2006, licensed under
      Creative Commons Attribution 2.

5.7   Spicosa team at study site 10, licensed under Creative Commons attribution 2.

List of tables

1.1 Some European Environmental Directives relevant to the Coastal Zone.   16

3.1 Properties of systems   54

3.2 IPCC 'storylines and scenario families', from IPCC-TGICA (2007).   71

3.3 Scenarios of the Millennium Ecosystem Assessment.   72

3.4 MEA categorisation of ecosystem services. From MEA (2005).   73

5.1 Stakeholder-issue mapping. Several ways to look at the stakeholder
    groups relating to the policy Issue of eutrophication in Himmerfjärden.   119

5.2 Himmerfjärden, south of Stockholm. An example policy Issue.   121

5.3 Mar Piccolo, Taranto: List of the main Human Activities,
    Dysfunctions and Impacts.   124

6.1 Regional and Global Institutions concerned with the
    coastal zone or SAF-like methods.   146

Biographical information

Denis Bailly (author) has been a lecturer in economics at the University of West Brittany (UBO) in France since 1993 and researcher at AMURE, the research centre in Law and Economics of the Sea at the European Institute for Marine Studies (IUEM) in Brest. He previously worked as researcher at the Maritime Economic Department in Ifremer, *l'Institut français de recherche pour l'exploitation de la mer*. His work has been mainly in the areas of fisheries and aquaculture economics, aquaculture and environmental sustainability, science and policy integration for Integrated Coastal Zone Management. He was joint scientific co-coordinator of the SPICOSA project that tested the theory in this book.

Marta Estrada (author) is Research Professor of the *Consejo Superior de Investigaciones Científicas* (CSIC) and a member of the staff of the *Institut de Ciències del Mar* (ICM), in Barcelona, Spain. She has more than 30 years of experience in biological oceanography and has published over 150 scientific papers. Her career has focused on marine phytoplankton ecology and has taken her on oceanographic cruises in the Mediterranean Sea and all the major Oceans. She has served in numerous national and international panels, including the Steering Committee (1999–2003) of GEOHAB (Global Ecology and Oceanography of Harmful Algal Blooms), an international programme of SCOR-IOC.

Tom Hopkins (author) is a Professor Emeritus of Oceanography from North Carolina State University, where he taught Oceanography, Global Problems, Environmental Science, and Simulation Modelling. He has enjoyed a long and varied career in research and teaching studying marine systems on both coasts of the United States, the Arctic and the Mediterranean and Black Seas. Over this career, his interests have trended from specific descriptions of marine environments to the complex systems that involve the interactions of humans and nature. He was joint scientific coordinator of the SPICOSA project.

Anne Mette (editor and author) is a research fellow at the *Kolleg für Management unde Gestaltung nachhaltiger Entwicklung gGmbH* (KMGNE) in Berlin, Germany. She is a social scientist focusing on the communication of complexity, and knowledge transfer strategies, at the interfaces of science and policy as well as those of science and society. Her special interest lies in transdisciplinary approaches with regard to climate change adaptation and mitigation strategies and vulnerabilities of coastal zones in developing countries. She led the group preparing the website at *www.coastal-saf.eu*, which expands on the material in this book.

Maurizio Ribera d'Alcalà (author) is director of research at Stazione Zoologica Anton Dohrn, Napoli, Italy, and leader of the Physical–Biological Coupling research group. He taught biological oceanography at the University Parthenope in Naples from 1995 through 2007 and is currently member of the Scientific Advisory Board in the Center of Excellence for the Analysis and Monitoring of Environmental Risk, in Naples, as expert for the Vulnerability of Coastal Systems. His research interests focus on biogeochemical processes and plankton dynamics in open ocean and coastal systems, and their response to climatic and anthropogenic forcing.

Audun Sandberg (editor and author) is a Professor in Social Science at the Faculty of Social Science, University of Nordland in Northern Norway. He has previously been Research Fellow at University of Dar es Salaam and University of Oslo and has been working closely with the Workshop in Political Theory and Policy Analysis at Indiana University on developing Institutional Analysis of resource governance. He has also been chairman of various committees in the Norwegian UNESCO Commission, in the Norwegian Research Council, in Norwegian Energy Industry and in the Norwegian Trekking Association. He has extensive research experience from the field of environment and sustainable development in Africa, Asia and in the European Union Area. He has recently been working on issues related to institutions for sustainable use of resources in northern areas, in particular in coastal and protected areas.

Paul Tett (editor and author) was Professor of Biological Oceanography in Napier University, Edinburgh until 2010, and is now Principal Investigator in Coastal Ecosystems Modelling at the SAMS Scottish Marine Institute, Oban, Scotland. He has worked in universities in China, France, Spain, Scotland, Virginia and Wales and carried out research on phytoplankton and eutrophication in coastal and oceanic waters of the northern Atlantic, publishing more than a hundred scientific papers and academic book chapters. Until recently he taught marine biology, scientific methods, and EU water quality legislation, to undergraduate and Masters students. He has served on the scientific steering committees of large UK and EU research projects and provided advice to the UK Government and the European Commission.

Editors' Preface

This book is about sustaining coastal zones as places for human habitation and enjoyment. To do this, we argue, it is best to see them as social-ecological systems in which it is necessary to protect the natural component, 'the environment', in order that its ecosystems continue to supply services for people. We describe a 'Systems Approach Framework', which has two main components. The first is the engagement of stakeholders at the interface between scientific knowledge and coastal environmental policy. The second is the use of models to simulate scenarios for alternate futures that can be used to evaluate policy changes and options for management. We provide examples of methods and results from the recent European project, SPICOSA (Science and Policy Integration for Coastal Systems Assessment), which brought together social and natural scientists and resulted in an unusual, transdisciplinary synthesis.

Metaphors are important in this book: metaphors of journeys and voyages; the formalised metaphors of conceptual models. So it is not inappropriate to introduce our argument in terms of coastal paths and sea-lanes. From high on a rocky headland there can be seen, spread out, the whole of the frontier zone between inland and the open sea. Our method is sometimes to draw back, to look at this broad view, and sometimes to zoom in on particular aspects, to enter workshops in which computer programmers sit before their terminals, capturing the essence of the natural environment with software. We will use the metaphor of ship-steering to describe the business of governing human societies and regulating human–environment interaction. This book itself may be likened to a vessel, freighted with a cargo of ideas, so that we are inviting you to take a metaphorical voyage in which you will meet with diverse shipmates: the philosophers Jürgen Habermas, Nicklas Luhmann, and Karl Popper; the ecologists Eugene and Howard Odum, the biologist and founder of general systems theory, Ludwig van Bertalanffy, the management scientist and developer of soft systems methodology, Peter Checkland, the ecological economist Robert Costanza, the institutional analyst, Elinor Ostrom, winner of a Nobel memorial prize in economics, and the environmental ethicist Holmes Rolston III. Back on land,

and still travelling with these ideas, you will encounter migrating salmon and little puffins, together with fishing skippers, perplexed city mayors, and nervous scientists. During this journey, you will be retracing the path taken by the authors and editors of this book, after being brought together in the SPICOSA project between 2007 and 2011.

This book tells you what we learnt along the way. It is a simple introduction to the transdisciplinary synthesis, which is described in detail at: *www.coastal-saf. eu*. In our text, we have drawn on technical terms from across the disciplines: from the 'soft' social sciences and the 'hard' natural sciences; and although we have tried to use a minimum of such words, we have not been able to avoid using some, and sometimes we have had to redefine these to suit our approach. Such key words, printed in **bold blue type on first** (substantive) use, are briefly explained in the glossary at the end of the book.

Acknowledgements

As editors, we acknowledge the help of other members of the editorial team: Marta Estrada and Maurizio Ribeira d'Alconà, our *coscienza critica;* and Tom Hopkins and Denis Bailly, our project co-ordinators, who designed and launched the good ship SPICOSA and steered it through often turbulent seas of theory and practice. Many other members of the project helped write the guidance material used in applications of the Systems Approach Framework, or SAF, and on which we have drawn for this book. They are listed at www.coastal-saf.eu. Without the many people who laboured to implement the SAF at SPICOSA study sites, we would have no evidence to report. Many of them are named as authors of papers in a special edition of the journal Ecology and Society, called 'A Systems Approach for Sustainable Development in Coastal Zones'. Finally, we acknowledge the European Community who part-funded SPICOSA as project 036992 under the 6th Framework Programme, and national programmes that completed the funding, between 2007 and 2011.

CHAPTER 1

Introduction

Paul Tett and Audun Sandberg

On the beach

Let's start on a beach: a Mediterranean beach. It is summer, and the sun is hot. People have come from the nearby city, to spend the afternoon here, or have flown in, on vacation from colder northern lands. They bask in the light and heat, while their children play on the golden sand and splash in the blue sea. The sparkling water's edge oscillates gently, offering and retrieving bits of seaweed, blades of eel-grass. Out on the sea, sailing dinghies pick up the light breeze, and small motor craft seek fish for the city's restaurants. Nearer the calm horizon, larger cargo vessels and ferry boats make their way to or from the city's harbour, part of an oceanic network of trade and communication.

Yes, this is an attractive scene, repeated around the modern world wherever people have leisure and money to travel. But all is not well. Behind this beach is a modern city. When it rains heavily, the city's sewers overflow and pollute the sea and shellfish with harmful bacteria. Nutrients in sewage, and fertiliser in rivers draining adjacent farmland, can enrich the sea and may cause *red tides* of algae that poison bathers or shellfish. Overfishing has depleted stocks. The beach itself is artificial, and has to be maintained against water movements; stabilising meadows of sea-grass that lay offshore have been destroyed. The city's mayor has to make choices. Expanding the harbour facilities will provide well-paid work for the city's voters, but will further damage natural coastal ecosystems. Increasing tourism will benefit hoteliers and restaurateurs but will add to sewage loads and will clash with port development. How does the city government make the right decisions about planning and public investment to ensure that the city remains sustainable? How does it ensure that citizens can be housed and employed in a coastal environment that continues to provide the **goods and services** of fish, shellfish, bathing water, and clean beaches?

This book is about methods to support such decision-making, methods that were developed and tested in a European research project, but which, we will argue, are of global applicability. The methods are drawn from a variety of disciplines, especially ecology, sociology and economics, and the synthetic discipline of systems theory. The rest of this chapter considers the nature and global importance of the coastal zone – the 'long narrow interface between land and ocean that is a dynamic area of natural change and of increasing human use' – and introduces some of the key ideas needed for the sustainable management of coastal systems.

Homo sapiens littoralis

The oldest human remains that are anatomically modern are about 200 millennia old. Our kind seems to have been, at first, rather unsuccessful, one species amongst many types of mammals inhabiting the East Africa savannah. At some time between 80 and 70 millennia ago, environmental change brought *Homo sapiens* to the edge of extinction, numbers falling to a few thousand. Somehow our ancestors survived this crisis, and, perhaps as a result of its strong selective pressures, grew fruitful and multiplied across all the continents of the world, until in 2010 we numbered 6.8 billion.

Evidence of how our ancestors spread is coming to light in studies of human genetics as well as of human remains and artefacts, and it begins to seem that the coastal zone – the seashore and the adjacent land and sea – provided the route for many migrations. It did this as much because of its wealth of resources – food from the sea and land – as because it offered easier passage. Until the development of agriculture allowed inland settlement, the densest human populations probably lived near the sea. And now that industrialisation has gathered the majority of humans into cities, the old preference, for the sound or the smell of the ocean, appears to be re-asserting itself.

The United Nations Environment Programme estimated that, in 2004, about 3 billion humans lived within 200 miles of the sea. That is to say, nearly half of our species lives in the 10% of land defined by the **LOICZ** project as 'the coastal zone', at a mean population density of more than two and a half times the terrestrial average. The US **NOAA** reported that in 2003, about 53% of the US population of 291 million lived in the coastal counties (including those bordering the Great Lakes) that made up only 17% of its land area. Of the United States' ten most populous metropolitan areas in 2000, nine (with 78 million people) were within 100 km of the sea or a Great Lake. But even the largest of these, the New York metropolitan area, with 21 million people on a peninsula and an island between the Atlantic Ocean and the Hudson river, only comes third in the global list of cities. First is Tokyo, with 35 million around Tokyo Bay; other coastal cities include: Jakarta (22 million, 2nd in the global list); Mumbai (20 million, 6th); Shanghai (15 million, 14th), Buenos Aires (12 million, 17th), Istanbul (11 million, 20th), Lagos (9 million, 26th) and London (8 million, 27th). These large cities grew up on fertile coastal plains, often close to navigable rivers, connected to the sea but sheltered from its worst storms, and sited for defence against raiders or invaders.

Rewards and risks

Individually and collectively, people like living in the coastal zone: *Homo sapiens littoralis* might not be a bad name for our subspecies. In the same way as our ancestors' long evolution in African savannahs may have given us a fondness for park-like landscapes, it may be that there is something in our inheritance that responds to the sight and sound of the sea. But of course, there must have been advantages to life near the shore in order to select for an inbuilt human preference, and there must continue to be benefits in building modern cities at the edge of the ocean.

In prehistory, the rewards of life in the coastal zone are likely to have been the wealth of resources there, with the potential to support large populations of humans. Archaeologists have found huge middens of shells, testifying to early human use of snails, limpets and bivalve molluscs from the seashore. Until modern times, the inhabitants of coastal north-western America feasted regularly on salmon returning up-river from the Pacific Ocean. The common feature here is a reliable supply of a protein-rich food, naturally concentrated at a fixed site from wider marine resources.

Modern coastal cities have outgrown such local harvests. They are giant vortices, sucking in food, water, energy and raw materials from all over the world, and the sea provides the main transport route for bulk goods. But while the network of roads, railways, sea-lanes and air corridors that feed people and materials into cities like our Mediterranean example is highly visible, the vortex involves two other, less obvious, networks. The first is economic: the inhabitants of a city must generate and export the wealth needed to pay for their imports. The second is material. Amongst other things, food and drink goes into the vortex, and much of it comes out again, transformed into wastes: the bodily wastes of humans and domesticated animals, uneaten food, etc. The sea is (in most cases) good at flushing away such wastes. Until early modern times, a city such as London relied on the ebb and flood of the tide to purge the Thames, its main river, of human excrement, dead bodies, decaying vegetables, and similar.

This didn't always work. To take just one example, sittings of the British parliament had to be suspended on hot summer days in the nineteenth century because of the stink from the river Thames. As cities have grown larger, their ability to overwhelm the waste-assimilative capacity of their local environment has increased.

Clearly, then, there are risks to dense human settlement in the coastal zone. The hazards are of two sorts. One type is generated internally, as we have just seen for London. Human societies tend not only to overload the capacity of local ecosystems to absorb human wastes, but also to exhaust the ability of these ecosystems to provide the goods and services that originally attracted humans there. The second type of hazard is generated externally. For example, 60% of Netherlanders live below sea level. They are at risk from flooding, as happened in 1953 when a storm surge in the North Sea overran the sea-dykes and killed 1,835 people (Delta Works, 2004). This was a natural event, but human-generated and other climate change increase the risk of further floods as sea level rises. The world is still emerging from a glaciation during which sea levels were tens of metres lower than at present. Stone-Age communities lived on what is now the Dogger Bank in the North Sea, until it was submerged about 8,000 years ago. Accelerated melting of ice that is presently accumulated on land may inundate more human-settled coastal regions during the twenty-first century. In contrast, a few people are lucky enough to live where the land is rising following the removal of the weight of ice-sheets, as around the Gulf of Bothnia in the Baltic Sea.

Natural and artificial

It is easy for the inhabitants of a modern coastal city to forget how artificial is the environment in which they live. Walk south along Strandweg ('Beach Road') in Scheveningen, one of the districts of Den Haag in the Netherlands. To the right are a few hundred metres of sandy beach between the concrete sea wall and the grey waves of the North Sea. For most people, that defines the width of the coastal zone, because, on the other side of the road, is the start of the city: the hotels and apartment buildings that form the western edge of a conurbation that houses 7.5 million people and extends northwards to Amsterdam and south-east to Rotterdam.

But most of the land on which these cities stand is man-made, a result of embanking and draining coastal marshes. A thousand years ago this region was, as apparent from surviving natural areas on the Dutch northern coast, a mixture of 'tidal channels, sandy shoals, sea-grass meadows, mussel beds, sandbars, mudflats, salt marshes, estuaries, beaches and dunes'. It was the place where land, river and sea

1.1 The shore at Scheveningen in the 17th century. Het Strand van Scheveningen, by Adriaen van de Velde, 1658.

interacted, salt tides penetrating tens of kilometres inland. Sediments stripped from the land during the Ice Age, and transported by glaciers, or more recently swept into rivers by rain and flushed down estuaries, were lifted and moved and re-deposited by waves and tides. Blowing sand accumulated in dunes, which were stabilised by grasses and shrubs until overwhelmed by storm-driven marine surges or by post-glacial rises in sea level. Birds flew in to feed on shellfish exposed on tidal banks; shoals of little fish shimmered and fed in shallow waters, swimming offshore as they grew. Scheveningen was first heard of in 1280 as a village on one of the larger dunes that formed islands of stability at the western edge of this salty, silty chaos. A painting by Adriaen van de Velde in 1658 shows the dunes, the beach, and the flat-bottomed boats used by the Scheveningers to get about and to catch fish, their main source of food and income. This painting summarises some of the key features of the coastal zone: the land–sea interface; natural and human-imposed change.

The coastal zone also includes coastal sea. Let's suppose that you board a ship offshore from Scheveningen, and that the ship begins to sail northwards at 12 knots, about 20 kilometres per hour. At first navigation is difficult amidst shoals and sub-merged banks, and the ship's motion rendered irregular by the short sharp waves in these shallow waters. If you are not too distracted by sickness to look at the sea, you'll notice that it is turbid, full of silt suspended in tidal streams. A day later, the water will be clearer and greener, the waves longer, the ship's motion easier, and the echo-sounder will show the seabed at 80 to 100 metres. After another day you will have passed between the Orkney and Shetland archipelagos and the echo-sounder will show that the ship is now in the North Atlantic Ocean, floating several thousand metres above the ocean depths. During the spring and summer you might see an abundance of gulls and kittiwakes here, or gannets diving: birds that have flown out from their nests on sea-cliffs or offshore islands. This is the outer edge of the coastal zone, the termination of the gradient from fully terrestrial to fully oceanic conditions.

Just as the landwards parts of the coastal zone have been remade by humans, so has the marine part become industrialised. This started with fisheries, of course, but in the first decade of the twenty-first century, some regions of the North Sea are dense with farms of giant wind turbines, or platforms for the extraction of gas or oil. Electric cables and pipes for gas or oil connect these offshore installations to the land. Industrial-scale fishing has ploughed the seabed and removed almost all the large fish, reducing the harvest and shifting the species composition towards smaller, more rapidly-growing types.

Many sheltered coastal waters are now home to aquaculture, growing fin-fish in floating cages, or shellfish or seaweed on platforms, buoyed lines or the seabed. In some east Asian bays, these operations occupy many square kilometres. And around almost all the world's coasts can be seen modern humanity's ubiquitous footprint: a litter or jetsam of plastic objects and fragments.

The value of the coastal zone

The North Sea and the low countries of northern Europe are not the only kinds of coastal zone that will be described in this book. Let us go now to a warmer climate, to a coastline fringed by coral reefs. What is the value of a reef? Without the coral animals and the encrusting seaweeds that make the physical structure of the reef, the shoals of brightly coloured fish would have nowhere to feed or to hide from predators. The limestone that is biologically deposited to make the reef is part of the natural cycles of carbon and calcium. So the reef has some sort of intrinsic value to the organisms that inhabit it, and to the planet as a whole. In this book, however, we will be concerned with the contribution of the coastal zone to human well-being, and thus, in this instance, with the value of the reef to humans. It has a use-value:

people can take its fish, for example. The exchange-value of the fish is what it would fetch if sold in a local market, or the cost of a licence for a tourist to hunt the fish under water. In the modern world, monetary exchange-value is convenient as a way of totalling lots of different use-values, and environmental economists have during the last two decades tried in this way to evaluate what nature provides to human society.

Robert Costanza and colleagues (1997) made such an estimate of the total global value of nature to humans. Their figure was 33 trillion US dollars per year at 1994 values. This can be compared with the global gross product of about 18 trillion dollars per year at the same time. The latter is the sum of money exchanges within human society (what people pay other humans for goods and services), and overlaps with the use-value only in so far as fishermen, for instance, are paid for fish. Most of the natural goods and services accrue to humans without payment, as exemplified by the service provided by a barrier coral reef in protecting shores against tsunamis. From Costanza's data it is possible to estimate a value for the coastal zone – or at least the part of it that comprises shallow coastal seas, lagoons, fjords, rias and estuaries, or is occupied by coral reefs, meadows of sea-grass, beds of seaweed, salt marshes, mangroves, and similar. This part of the sea occupies only 6% of the area of the planet's sea and land, but is 43% of the global worth of natural services.

There is also the value of the human assets in the coastal zone. As an example, New Orleans stands at 1165th in the list of the world's most populous cities. Nevertheless, in 2006, a year after its flooding as a result of hurricane Katrina, insurers and the US federal government were estimating restoration costs of $200 billion. The official death toll, of 1,464, although about 200 times smaller than lives lost to the tsunami in the Indian Ocean in 2004, was deemed shocking for an advanced industrial country such as the United States. In addition to the loss of lives and physical infrastructure, there was a further price to be paid. Most of the pre-flood population of 485 thousand left the city in the week following the disaster. Less than half had returned by the following year. Hundreds of thousands of people had to find new homes, new jobs, new schools for their children; an enormous amount of **social capital** was destroyed and had to be rebuilt. Although economists don't put monetary values on it, such social capital – the local networks of human trust and mutual support that may be diagnostic of *Homo sapiens societatis* – is part of the wealth of coastal zones.

Sustainability

The millionfold increase in *Homo sapiens'* numbers – from a few thousand 70 millennia ago to 6.8 billion in 2010 – has impacted on non-human life more, and more irreversibly, than the preceding glaciation. The increase needed only about 22 doublings of the initial population: such is the inexorable mathematical logic of exponential growth. At first, doubling was slow, each taking about 3 millennia or about 130 human generations. Growth quickened during recent millennia, and the last two doublings took only about 50 years each.

Three things led to this enhanced multiplication. First, human ability to make extensive social networks and to depersonalise these as institutions of production and government. Second, evidence-based technologies for growing and distributing food and for ensuring public health. Third, our increased use of energy. During the last hundred years, exploitation of fossil fuels – coal, oil and gas laid down in rocks over hundreds of millions of years – has allowed the average human in the year 2000 to command about 20 times as much energy as is generated by personal metabolism: it is as if each of us has 19 slaves to do our bidding. We use this energy to heat (or cool) our homes, power our cars and transport systems, and, critically, to make fertilisers and pump water. Without the crops thus nourished, many people would starve. Without the energy to purify drinking water and remove sewage, many humans would die of infectious diseases.

Thus our planet is now home to more people, using more energy and resources, than ever before. The consequent impact on the world is well known. Farmlands and managed forests, power stations and cities, roads and harbours and playgrounds, have spread over and denatured ecosystems like mould on bread. Humanity's **ecological footprint** is now several times larger than our planet's land area, our burning of fossil fuel is changing the Earth's climate; and it is quite unlikely that there are sufficient global resources to support the present population – far less the 9 billion projected for 2040 – at the standard of living presently enjoyed in Europe and North America. Indeed, the amount of natural resource available to the average human is halving every two decades. *Homo sapiens'* first existential crisis took place at the start of prehistory. It was how to survive at all, as drought intensified in north-east Africa. Now, having grown fruitful and multiplied, our species is faced with its second existential crisis – that of how to live sustainably in a finite world.

Because of its large population, the coastal zone is the most important part of the human habitat exposed to sea-level rise and other consequences of global climate change. In addition, its enhanced population density intensifies local human impact.

Not only are there more people per hectare, but their demands for food, water and other resources cannot, in most cases, be satisfied from within the zone, but must be shipped or trucked from the rest of the world. As already mentioned, the cities of the coastal zone are vortices that suck in enormous quantities of supplies each day, and potentially discharge enormous quantities of waste into the local environment.

Beyond the cities are coastal zone regions where people still live in comparatively low densities – for example, around the fjords that penetrate the Atlantic coasts of Scotland and Norway – making, until recently, much of their living from the sea. Here, one might think, it would be easier to live in greater harmony with nature. But in fact these remoter communities are also drawn into global networks of production and consumption. As an example, industrial-scale aquaculture provides good jobs for local people, but impacts on the environment: a typical salmon from one fish-farm is equal to that of a mid-size human city and consumes hundreds of tonnes of feed each month. The feed is made from small fish, caught in distant oceans. The excreta can disturb the local environment. The fish themselves are exported to supply the food needs of city-dwellers across the planet. The farming of salmon is perceived as being harmful to wild populations of the fish, and this leads to conflict between aquacultural interests and those of landowners (who gain income from recreational fishing for salmon).

Clearly there are, here and in the other examples we've given, both local and global issues relating to sustainability. In the case of salmon-farming, the global issues are exemplified by the way increasing human numbers and food needs have combined with exhaustion of many wild fish stocks, to make aquaculture seem a rational alternative to commercial fisheries, and by the problems in finding food for the salmon. The local issues concern the environmental impact of farms, including their effects on wild salmon.

We are, in this book, largely concerned with such local issues. There are several reasons for this focus. It is in accord with the slogan: 'think global, act local', which includes the idea that the answer to the big questions might be assembled from the answers to many small questions. Local problems are often more tractable than global problems, because local responses can be faster, and local solutions easier and cheaper to implement. In the case studies reported here, we were mainly constrained by the amount of funding to problems that could be tackled by small groups of scientists. Nevertheless, finding methods that solve local, small-scale, problems increases confidence that when we scale these methods to the global problems, they will do more good than harm.

Humans and the environment

An ecological system, or **ecosystem**, is 'any area of nature that includes living organisms and non-living substances interacting to produce an exchange of materials between the living and non-living parts' (Odum, 1959). A salt-water pool amidst rocks on the seashore, containing seaweed and marine snails, is an example of an ecosystem; so is the much larger and more complex system of the European North Sea. Such systems have a biotic part, a community of living creatures, interacting with each other in a food web, and an abiotic (non-living or physico-chemical) part including a source of energy such as sunlight, and substrates or media such as water, air, soil, or rocks.

Many ecologists prefer to see humans as part of the biota, and thus deem people's physical existence to be fully within ecosystems as defined above. After all, men and women get food from plants and other animals, breathe the same air as other creatures, and add wastes to the common stream. Such inclusivity runs into two objections. One is that treating human life as part of nature leaves undefined the position of human institutions such as law courts and libraries, markets and manufactories, which have no counterpart in the life of any other species. The other arises from human conceit. Many people think of themselves as uniquely different from other animals, and most legal systems have traditionally given more rights to humans than to animals, plants or ecosystems.

According to this anthropocentric point of view, natural ecosystems are part of the natural environment, which is distinct from the world of human artefacts. This distinction is captured in Figure 1.2, which shows humanity's slice being withdrawn

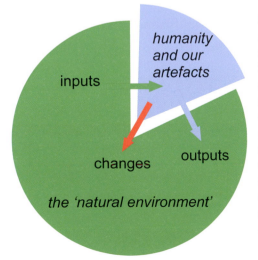

1.2 Humanity becoming distinct from natural ecosystems. The inputs from nature to the human population are food and raw materials, and the outputs are wastes from humans and their activities. Changes include obvious effects of human activities, such as clearing woodland to make farmland, and unintended effects such as those caused by fishing (an *input*) or pollution (an *output*). Some ecologists use the term 'pressures' to refer to the human activities that have the potential to cause changes in ecosystems.

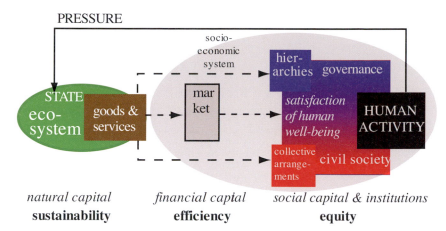

1.3 A social-ecological system. Its components are the human (socio-economic) system and the (natural) ecosystem. The two parts are shown here as distinct, with more emphasis on the human part. Human interactions with nature are seen in terms of goods and services acquired and pressures exerted. Each part of the system contains a particular form of capital and associates with a particular goal.

from nature's pie. Although this is only one way to conceptualise the world, it will be useful to see things in that way, at least temporarily, in order to investigate interactions between *Homo sapiens* and nature in more detail. And it leads to Figure 1.3, which provides a social and economic framework for understanding these interactions.

The key idea in Figure 1.3 is the recognition that humans have needs, that the satisfaction of these needs increases human well-being, and that much of human behaviour is motivated by the pursuit of such satisfaction. Environmental economists use the label **ecosystem goods and services** for what humans take from nature to satisfy these needs, the goods including fish from the sea and the services exemplified by the capacity of marine ecosystems to recycle human wastes. Of course, a fish has a value to itself, and a stock of fish has an intrinsic value to its species and to the ecosystem of which it forms a part (Rolston, 1994). But as we move our conceptual focus from ecosystem to socio-economic system, the value of these things becomes instrumental; we appreciate them for their utility to humans.

Figure 1.3 aims to show the three ways in which these benefits are brought into and distributed within human societies. In addition it signals the adverse effects that activities carried out by humans, in getting and using these goods and services, can have on the ecosystems that provide them. The three routes or arrangements are called collective, market, and hierarchical. Each involves sets of rules and the institutions and organisations that embody these rules. For example, a challenge for a society with important fisheries is to find a way to sustain these fisheries: to avoid situations in which rational individual decisions lead to collective over-exploitation. *Collective arrangements* are the rules made by fishermen themselves. Often the result of long experience and 'traditional ecological knowledge', these arrangements run into difficulty in times of new technologies or external interventions.

Market arrangements are supposedly impersonal. Markets as physical places for trading goods, services and information have been known for thousands of years. Currency facilitates such trades, and the **market** is nowadays understood as an impersonal institution that, so economists claim, ensures that the supply of goods and the demand for them is put into balance through adjustments to prices. This process is what Adam Smith called 'the invisible hand', which maximises human welfare by aggregating many small actions arising from self-interest. In principle, the market can automatically control fisheries, for as stocks dwindle, prices rise, and people should be less willing to buy fish. In practice, such automatic regulation often fails; it has become clear that markets need externally imposed rules to work properly. These can be imposed by *hierarchical arrangements*, involving a chain downwards from rulers to ruled that is underpinned by law and power, and a chain upwards from ruled to rulers that depends on custom and perceived legitimacy. In most modern societies, ruling is better seen as a function of institutions, such as Fisheries Ministries, rather than of princes or presidents: as will be explored in the next section under the heading of **governance**.

The use of a collective, a market or a hierarchical arrangement, or a mixture, depends on the kinds of ecosystem services to be distributed, and the administrative resources and political histories of the societies concerned. For example, the Baltic Sea has provided humans with codfish and herring for millennia. The national archives of Denmark, Estonia, Finland, Latvia, Russia and Sweden, which record taxation of the fishery, show how landings varied between 1550 and 1860 (MacKenzie *et al.*, 2007). These early modern states may have provided protection for fishermen and markets, but there is no evidence that they tried to regulate catches, and indeed overfishing is unlikely to have been a problem when boats were powered by wind and nets hauled by human muscle. It may be imagined that skippers and crew shared information about where to fish, perhaps keeping secret only the sites of their very best catches.

Fisheries technology and practices changed after 1880. '[B]oats became motorised, hydraulic winches were implemented, off-shore fishing grounds started to be used and nets and boats became larger' (MacKenzie *et al.*, 2007). Centralised decision-making began to influence fishing. For example, during much of the second half of the twentieth century, the command economy of the USSR managed the small Baltic state of Latvia as the base for a large deep-water fleet and a processing industry to provide for fish for much of the Soviet Union. Following Latvia's independence in 1991, most of the high seas effort was abandoned, and the fishery and processing industry privatised to operate in a market economy that relied to a much greater extent on local fisheries. 'Distribution of fish on the domestic market is direct, with little wholesale activity. Smaller processors and traders buy fish directly from independent Baltic and coastal fishermen at the landing point. Prices are determined by direct negotiation, payment is in cash, and vary according to seasonal supply and demand.' (Megapesca, 1997). Since accession to the European Union in 2004, the fishery has been subject to the EU's Common Fisheries Policy, which currently requires member states to control fisheries effort in order to protect stocks.

Now let's return to Figure 1.3. In 1998, Fikret Berkes and Carl Folke coined the term **social-ecological system** to refer to 'linked systems of people and nature'. They argued that 'humans must be seen as a part of, not apart from, nature — that the delineation between social and ecological systems is artificial and arbitrary' (Stockholm Resilience Centre, 2010). Figure 1.3 shows just such a social-ecological system, and understanding and managing it requires the joint efforts of three academic disciplines, each with a different way of looking at the world. Economics, broadly speaking, deals with the way humans produce and use resources to satisfy well-being needs. Sociology describes and analyses human social activity and institutions. Ecology seeks to understand the role of living beings in the workings of the natural world. Thus the management of a social-ecological system should aim at finding solutions that are simultaneously ecologically **sustainable**, economically **efficient**, and socially **equitable**. The need for sustainability should be obvious. The fair sharing of ecosystem goods and services is not only desirable for ethical reasons, but also because it is becoming clear that inequitable societies are both less happy and less able to become environmentally sustainable (Wilkinson & Pickett, 2009). Finally, economic efficiency is needed to get the greatest benefit from these finite natural resources.

Governance (*Homo sapiens societatis*)

Humans are social creatures: we are good at *collective rationality* at least as much as individual rationality, although, as history shows, not all group decisions have proven wise. What seems to distinguish *H. sapiens* from other species of the genus *Homo* is an ability to organise people not just by dozens, but by millions. Broadly speaking, this is done through establishing **institutions**. These are, in essence, sets of rules – later on, we will see them as evolving systems – such as those that organise religious activities into churches or educational activities into schools. Property rights are fundamental institutions that link the social sphere to the bio-physical sphere, providing a basis for the existence of money and markets. A human society can thus be seen as a set of overlapping, nested and interlocking institutions that provide a particular solution to the eternal dilemma of collective rationality: how to balance private autonomy and individual rights against the interests of society as a whole. Institutional development is therefore crucial in a modern society, and is to a large extent taken care of by its system of governance.

Whereas governing is what governments do, *Governance* is the whole chain of 'steering', leading upwards from the people giving authority, to the leaders that shall rule them as subjects or citizens of a State, or members of a voluntary organisation, and downwards from these rulers. The word derives from 'gubernator' in Latin and from the classical Greek κυβερνήτης, a ship's steersman, helmsman or pilot: the captain or navigator who sets a ship's course as well as the person holding the steering oar or tiller. As long as the crew and the passengers trust the steersman, his/her power is legitimate. The etymology suggests a metaphor: the need to steer the ship of humanity through stormy waters to reach a haven of prosperity and equity. From steering a people, to steering a nation, and now to steering the planet, with its cargo of 6.8 billion people, there is a long path of development.

Humanity is, collectively, becoming aware of the potential consequences, for human economies and societies, of climate change, overpopulation, and environmental damage. Leaders of most of the world's states met in Copenhagen in 2009, aiming to make an agreement to reduce carbon emissions and thus decrease the threat from global warming. But as the disappointing results of this conference illustrated, global governance is struggling to find effective, efficient and just remedies. Traditionally, governance has been associated with 'what governments do', which is to say: with political leaders and formal political institutions that are able to enforce decisions. Such authority and institutions exist, in the modern world, primarily at the level of states. Thus 'global governance' is something of a contradiction in terms. The United Nations is not a world government, but with its many committees and agencies acts as a debating chamber for sovereign states, agreeing on joint action typically only when this is in their interests. Nevertheless, much has been achieved in terms of influencing member states by UN agencies, international secretariats and transnational non-state actors. We will encounter some of these, such as LOICZ, the international research project dealing with Land–Ocean Interactions in the Coastal Zone (Crossland *et al.*, 2005), and the international Convention of Biological Diversity, adopted in Rio de Janeiro in 1992 and recognising in international law that the conservation of biodiversity is a common concern of humankind. This kind of multilevel, polycentric governance is, perhaps more typical of many systems of governance than readers might suspect, and one that can ensure both legitimacy and sustainability better than more totalitarian systems.

As we will use the word in this book, governance involves three functions. The first is making the rules for how to make policies (laws, regulations, guidelines, etc.). This is equivalent in the 'ship of humanity' frame of reference to the ship's construction principles and the rules for appointing captains. The second function is like a sailing plan, which is based on a collective decision of where we want to sail and who/what we want to bring along. The third function, the setting of the daily course and the tasks necessary to move the ship, is based on operational rules for good seamanship. We will refer to the first of these as the constitutional level of governance, to the second as the collective decision level, and to the third as the operational level.

Research shows that most provincial and local governments, with or without formal legislative authority, as well as self-organised groups in coastal areas, do indeed conduct their constitutional choices, their policy-making/collective decisions, and their current operations, in accordance with such a three-level system of governance (Ostrom, 2005). However, states sometimes make the constitutional choice of devolving the entire field of environmental policy-making to semi-autonomous regions, or they give up some of their sovereignty and agree on common policies with other states. Bringing the collective decisions closer to the people affected by them helps to increase the legitimacy of the policies, and thus to improve their implementation. Solving cross-border problems by cooperating with other nations helps to produce solutions that are better for more people.

The European Union, with 26 member states in 2010, has diffused power in both directions. For example, environmental health in the coastal zone is guaranteed by the Directives listed in Table 1.1. Such directives are currently issued by the elected European Parliament and the Council of Ministers appointed by state governments, under provisions of the treaties that established the Union. Member states are obliged to transpose these Directives into their own laws. By such incorporation, the overall aim of maintaining ecosystem quality can be made to work together with national,

Table 1.1 Some European Environmental Directives relevant to the Coastal Zone. Some have been updated since the initial date.

Directive (short name)	Initial date	Regulates/protects
Dangerous Substances	1976	regulates pollution of the aquatic environment
Bathing Waters	1976	protects recreational waters
Freshwater Fish	1978	protects quality of freshwaters with fisheries
Birds	1979	conserves certain bird species
Shellfish Water	1979	protects water quality for shellfisheries
Urban Waste Water Treatment	1991	requires treatment of sewage discharges
Nitrates	1991	reduces leakage of agricultural nitrates into natural waters
Habitats	1992	conserves certain species and habitats
Shellfish Hygiene	1996	prevents contamination of marketed shellfish
Water Framework	2000	protects quality of freshwaters and of the sea close to land
Marine Strategy Framework	2008	protects quality of salt waters

regional or local customary law. For example, the *European Water Framework Directive* (**WFD**) is implemented in Scotland (a semi-autonomous region within the United Kingdom) by a Water and Environmental Services (Scotland) Act, which gives powers to ministers of the Scottish Government to make regulations about water quality, taking advice from the Scottish Environment Protection Agency. The Agency is charged with policing these regulations, and is aided by planning decisions made by local governments in rural counties or urban municipalities.

Many other states have made laws and established agencies to protect environmental quality. For instance, legislation to prevent the harmful effects of nutrient waste on coastal seas includes: Australia's Oceans Policy; the Marine Environmental Protection and Prevention and Control of Water Pollution Laws of the People's Republic of China; and the United States' Oceans and Harmful Algae and Hypoxia

Research and Control Acts. But whereas a modern state has well-defined boundaries and aims to regulate its own economy and laws within these boundaries, a coastal zone is less well demarcated. Take the composite Mediterranean city with which we started this chapter. Its environmental policies will be set by its national or regional government's implementation of international agreements and EU directives, by national laws, and by interactions between the city government, national environmental agencies, and stakeholders who use local ecosystem goods and services. Conditions in its coastal waters can be influenced externally: by migratory birds and boats from across the sea; by currents bringing water from other parts of the Mediterranean; and by rivers arising upstream from the coastal zone. We will see that in order to address environmental problems in coastal zones, it will be necessary to define appropriate physical boundaries and to understand the larger international scale as well as the local institutions of governance.

Stakeholders

A **stakeholder** is a person who has a legitimate stake or interest in a particular issue. In the present case, the relevant issues are those associated with problems of sustainability, efficiency and equity in the coastal zone. 'Person' can mean a human being or a legal entity such as a commercial enterprise. 'Legitimate interest' could mean a legal claim or right, but, more generally, implies that the person's well-being is affected by the problem or its solution in a way that other people would understand and consider to fall within social norms even if they did not agree with the nature of the interest. For example, who has a legitimate interest in the pods of whales that migrate each summer up the western coast of Europe? Norwegians would like to hunt them; Scots would like to protect them (Scott & Parsons, 2005). In both Norway and Scotland, whale-watching industries have grown up. Both the whalers, and the skippers of the whale-watching boats, have a legitimate interest in the whales, even though the two sets of interests are somewhat opposed. There are ethicists who argue that whales also have interests, or are, at least, 'worthy of moral consideration' (Johnson, 1991), but, as was briefly discussed earlier, this book's position in relation to nature and non-human creatures is strictly instrumental: natural things have value in so far as they are valued by humans.

What about the tourists who go whale-watching, presumably because they are interested in whales? We would only say that they had an interest if their well-being depended on being able to see whales. If the trip is just a holiday, and could be substituted by a visit to the cinema, we don't count them as stakeholders in whale matters. A recent survey of visitors to the west coast of Scotland asked for their opinion of salmon farming: did the existence of the somewhat industrial infrastructure of floating cages and barges detract from their enjoyment of the scenery or their willingness to re-visit? (Nimmo *et al.*, 2009). Most thought that it did not, in part because they recognised that people's livelihoods depended on the farms, whereas they (the visitors) were only passing through. The point is that, whether or not people call themselves 'stakeholders', most can tell whether or not they have an interest that requires their voices to be heard.

A different problem is that of getting heard. Stakeholders have a moral claim to be heard in respect of their interests, but some voices may be drowned by others, more articulate or more powerful. Thus the management of a stakeholder engagement process needs some care, and can be helped by the tools we will describe in later chapters. Where governance is strong and reliable, it may be that public officials can oversee the process, themselves appointed by governments elected by citizens, and so explicitly responsible to the public. (Note the difference between the roles of stakeholder, with a moral claim in respect of an identified matter, and citizen, with a moral right to choose between potential alternative governments.) Where governance is distant, weak, or corrupt, stakeholder engagement may be the only way of getting things done.

The process whereby stakeholders jointly consider and evaluate information is called **deliberation**. It would seem to be no more than common sense to involve people in debating matters that impact on their lives and livelihoods. Our argument goes further, however, drawing in chapter 5 on the work of Jürgen Habermas, who claims that the ability to communicate with each other is a key result of human evolution and is thus built into our natures, and that, by talking sufficiently and seriously, people can sometimes reach a consensus about the natures of, and the solutions to, problems that affect their interests.

Systems

The world is made up of interconnected things. Consider, in the natural world, a puffin (a comical little seabird) catching a small fish: two things and a relationship. Many such meals take place in the coastal zone, and ecologists generalise them in terms of a food web, with boxes labelled 'seabirds', 'fish', 'plankton', etc., and arrows labelled 'feeds on'. In the human world we find other things that we think of as real, such as a person or a sum of money. These are not always real in the sense of physically existing, because what we mean by a particular person is, typically, not their body but their mind or personality; and what we mean by money is not the metal coins or paper notes but the value that these represent. Nevertheless, just as with natural-world things, we can generalise, or statistically aggregate, such human-world things by referring to 'human populations' (the total of bodies), 'public opinion' (a total of the expressed intentions or views of many persons), and 'economies'.

Is the world any more than lots of these things, big or small, simple or aggregate, real or intangible, connected together in a complex web? Systems theory argues that there exist patterns of relationships in such networks, and that the components and sub-nets are arranged into hierarchies. Any pattern of relationships constitutes a system: it may be a temporary aggregation of parts, which will soon dissolve; or it may include feedback loops which give the system some **emergent properties** that are not part of the system's components. A simple example is the combination of a domestic hot-water boiler, a pump, radiators, and a thermostat. The thermostat is set to switch on the boiler and pump when the house's temperature falls below a set value, and to turn the heating off when the temperature rises above that value. The result is that the house is maintained at a near-constant temperature. This temperature regulation is an emergent property of the system, a property that it is not to be found in any single component.

To be more precise, temperature regulation is an emergent property of a system that contains a **negative feedback** loop. Consider, next, a system made up of finely-ground charcoal, sulphur and saltpetre. You might recognise this mixture as gunpowder. Left alone, the system is inert. Heat a part: the mixture ignites locally. The saltpetre begins to oxidise the charcoal and the sulphur. This generates more heat, and thus the reaction spreads rapidly, involving more and more of the powder in this **positive feedback** loop. The resulting explosion of hot gases is an emergent property. In this case, it is short-lived, terminating when all the gunpowder is burnt. The contrast between such an explosion and temperature stabilisation points to the comparatively long-lived nature of systems with self-regulation (**homeostasis**, as it is often called).

According to a systems view, life itself is made up of homeostatic systems or is an emergent property of such systems. Organisms are the result of the interaction between the information contained in DNA and the DNA's environment, bounded on at least two levels – that of the cell and that of the body. Cells are systems within the larger system of the body. The environment that shapes organisms is itself part of an ecosystem within which the organisms live, and within which their DNA-information evolves. Human organisations can also be seen as systems, able to evolve and adapt as their social, economic and physical environment changes. In the case of systems that survive, such evolution is a homeostatic process. But we also know of human systems that, in effect, exploded and vanished, leaving behind only puzzling relics: giant statues half buried in shifting sands or on barren hillsides. In his book, *Collapse*, Jared Diamond (2005) provides examples of such a disappearance, his paradigmatic case being the initially successful human settlement of Easter Island in the Pacific Ocean. This led to population growth, environmental degradation and the downfall of a society that could not adapt to changed conditions and which had lost its ability to sail away.

It does seem that natural systems (ecosystems) have, as a result of natural selective processes acting on their constituent species, evolved emergent properties of self-regulation, enabling them to resist, or recover from, a certain amount of disturbance. When humans were few and the world seemingly infinite, it was usually easy for tribes to move on from regions of ecological damage (Mithen, 2003), leaving them to recover slowly. This is no longer possible in the crowded modern world, especially in the coastal zone. Thus there is a need to understand how natural ecosystems and human social systems fit together, so that management of environmental problems can make use of, or strengthen, stabilising interactions within and between them. Consequently, this book looks at how coastal zones can be understood and managed as social-ecological systems, as shown in Figure 3, and sees sustainability and resilience as properties of the combined systems.

A systems approach to a sustainable coastal zone

For many of our illustrations of a systems approach to the coastal zone, we draw on the results of a European research project that was operational between 2007 and 2011. SPICOSA – the acronym stands for 'Science and Policy Integration for Coastal System Assessment' – studied the 18 sites shown in Figure 1.4.

SPICOSA's method involves collaboration amongst three groups of actors: scientists, who have the technical skills to analyse and numerically model cause-and-effect chains from human activity to impact on ecosystem goods and services; environment managers, who are agents of government and are charged with planning new developments, the maintenance of environmental quality and the regulation of impact; and stakeholders, who have an interest in a relevant human activity or its impact. Earlier procedures for ameliorating environmental impact in European states were either 'top-down', controlled by managers, or involved dialogue between managers and scientists, and often failed to work. SPICOSA's

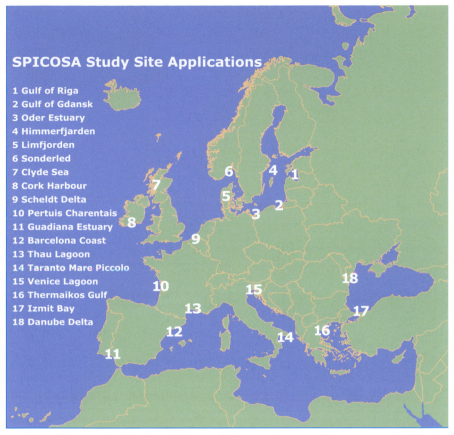

SPICOSA Study Site Applications

1 Gulf of Riga
2 Gulf of Gdansk
3 Oder Estuary
4 Himmerfjarden
5 Limfjorden
6 Sonderled
7 Clyde Sea
8 Cork Harbour
9 Scheldt Delta
10 Pertuis Charentais
11 Guadiana Estuary
12 Barcelona Coast
13 Thau Lagoon
14 Taranto Mare Piccolo
15 Venice Lagoon
16 Thermaikos Gulf
17 Izmit Bay
18 Danube Delta

1.4 Map of Europe showing the 18 sites at which coastal zone problems were studied during SPICOSA.

argument was that including stakeholders in the identification of a problem and in evaluating management choices or policy options would lead to improvements in the quality, effectiveness and legitimacy of public decisions about the coastal zone environment.

SPICOSA called its procedure the **Systems Approach Framework,** or **SAF**. This has the five steps shown in Figure 1.5. An application starts with **Issue Identification**, which is a process of focusing on a problem caused by a human activity that impacts on the supply of ecosystem goods and services, leading to an agreement amongst stakeholders and managers on the policy choices or management options for solving the problem. What we call the **Issue** is the identified problem plus the agreed set of possible solutions that are called **scenarios** because they may also need to take account of external pressures on the coastal zone. The second step, **System Design**, drafts a plan for a model system that includes the essential features

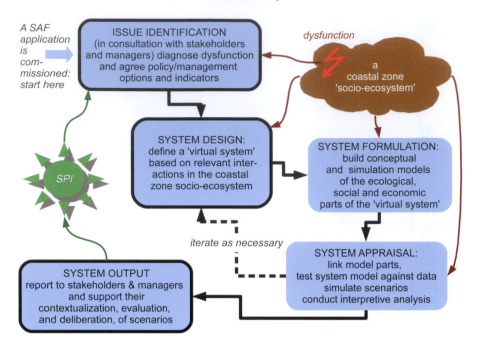

1.5 The steps of SPICOSA's Systems Approach Framework or SAF. The Issue is a dysfunction in the social-ecological system (including its economy) involving a human activity's impact on ecosystem goods and services. The symbol, SPI (Science–Policy Interface), corresponds to the similar symbol called 'communications space' in Figure 1.6.

of social-ecological system behaviour involved in the problem. It bridges the gap between the system in the real world and one or more simplified **conceptual models** that will be used as the basis for the next step. Step three, **System Formulation**, involves finding the appropriate mathematical formulae to transform a conceptual understanding of the interactions between system components into a quantitative **simulation model** that, having been validated, can be used to explore the different effects of the management or policy options. Step four, **System Appraisal**, carries out these simulations and subjects the quantitative results, and qualitative data, to ecological, economic and social appraisal by the specialists. Finally, step five, **System Output** reports back to managers and stakeholders, to inform them about the likely consequences of particular options and thus to aid stakeholder deliberation and management or government decision.

Figure 1.6 views the SAF in terms of the institutions and actors (or rôles) involved in it, and emphasises the key attributes of each: *Science* mobilises *knowledge* to explain the dynamics of selected coastal zone systems and to explore the potential consequences of alternative policy scenarios or management options; *Stakeholders*

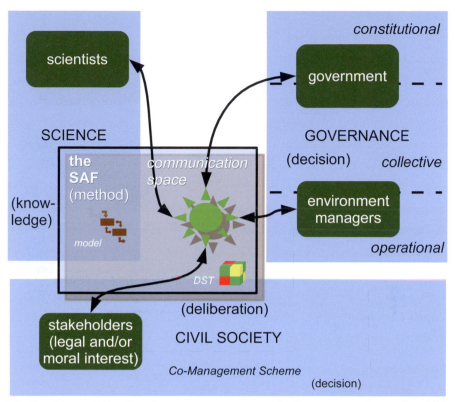

1.6 The Science–Policy Interface, seen as a communications space, a forum in which governance, civil society and science interact. Each of the large rectangles represents an institution; the smaller, rounded-corner, rectangle represent 'actors': groups or organisations of people operating according to the rules of these institutions. The parenthetical words refer to the main attribute of each institution (in relation to environmental problems). Thus the key attribute of Science is knowledge obtained according to defined procedures: there is, of course, also knowledge elsewhere, but it is not defining. Civil society is also shown to contain decision-making institutions for environmental co-management, which are part of the collective arrangements shown in Figure 1.3. SPICOSA's SAF is a set of rules (and, therefore, an institution) with two main functions: to open a space for communications between the actors (playing the rôles of stakeholder, scientist or government (official)/ environment manager); and to evaluate policy options or management choices relating to a socio-ecological Issue. The symbol used for the communications space corresponds to the symbol SPI in Figure 1.5.

deliberate on the basis of their interests and this knowledge; and *Governance* decides in the interests of society as a whole.

This book is about what we learnt during the SPICOSA project: about the SAF method itself; about coastal zone ecosystems, economies and societies; about how to scale up to provide advice to the constitutional level of governance; and about the possibilities for using the methodology elsewhere in the world's coastal zone.

Chapter 2 deals in greater detail with coastal zone problems, especially those identified at SPICOSA study sites. Chapter 3 explains systems theory and the systems approach as applied to the coastal zone. Chapter 4 presents the main tools available to natural scientists, social scientists, and economists, to analyse, describe and simulate the properties of coastal zone systems. Chapter 5 returns to the interface between the institutions of Science, Governance and Civil Society shown in Figure 1.6, and brings back into focus the stakeholders – the people and institutions on whose behalf the tools are used by the specialists. From an ethical perspective, of course, the stakeholders come first: they use the ecosystem goods and services of the coastal zone and suffer or benefit from solutions to coastal zone problems. We have already argued, and will seek to demonstrate, that involving stakeholders leads to better as well as more legitimate solutions. However, they are only a part of Civil Society and **collective rationality**, and so we will need to consider their proper roles in interaction with the operational and collective levels of Governance. Lastly, chapter 6 assesses the utility of the SAF at SPICOSA's European study sites and its potential for application in other parts of the world.

EndNotes

[On a beach] This Mediterranean city is a composite, and combines problems experienced by Barcelona, Marseille and Athens, amongst others. The definition of the coastal zone was taken from the *Science Plan and Implementation Strategy* 2005 of the international project LOICZ (Land–Ocean Interactions in the Coastal Zone, website in 2010 at: http://www.loicz.org/).

[*Homo sapiens littoralis*] Concerning human prehistory and migrations, *see* Macaulay *et al.* (2005), Mellars (2006), Oppenheimer (2004), and http://www.bradshawfoundation.com/stephenoppenheimer/index.html. The US Census Bureau (http://www.census.gov/ipc/www/popclockworld.html) gave an estimate of 6,890,944,643 for the global population at 10:27 universal time on 2 January 2011. The UNEP estimate of the coastal zone population comes by way of LOICZ and Crossland *et al.* (2005) from a primary source in Nicholls & Small (2002). The US coastal zone estimates are from Crossett (2005) and Hobbs & Stoops (2002). The coastal zone metropolitan areas (named by major city) are: New York, Los Angeles, Chicago, Washington, San Francisco, Philadelphia, Boston, Detroit, Houston. Dallas is the major city that does not lie close to the sea or a Great Lake. Estimates of populations of global cities (including wider metropolitan areas) were taken from http://www.mongabay.com/cities_pop_01.htm in January 2010, which cited 'Wikipedia sources'.

[Rewards and Risks] Concerning human attraction to the sea: Morgan (1982) proposed that humans part-evolved as water-living apes. Concerning middens, *see*: http://archaeology.about.com/od/boneandivory/a/shellmidden.htm and an example description by Melton & Nicholson (2004). Mithen (2003) reviews late prehistoric human–environment interactions. Details about nineteenth century London are given at http://www.victorianweb.org/science/health/thames1.html. Concerning evidence of post-glacial human occupation of lands now submerged beneath the North Sea, *see* Coles (2000) and Weninger *et al.* (2008).

[Natural and artificial] The Netherlands Randstad, or 'rim city', that includes Scheveningen is made up by 'the four largest Dutch cities (Amsterdam, Rotterdam, The Hague and Utrecht), and the surrounding areas' according to http://en.wikipedia.org/wiki/Randstad. The description of natural conditions in the Netherlands is based on UNESCO's account of conservation areas in the present-day Wadden Sea, at http://whc.unesco.org/en/list/1314. The description of sailing northwards from the southern coast of the North Sea to the outer limits of the continental shelf is based on the first authors' experience (1989–2010) during research cruises in, and ferry crossings of, the North Sea and north-western Scottish waters. Fisheries impacts in the North Sea are documented by Thurstan *et al.* (2010). The remarks about the extent of Asian aquaculture derive from an FAO expert workshop on 'Building an Ecosystem Approach to Aquaculture' in May 2007.

[The value of the coastal zone] Putting monetary values on nature, as exemplified by Costanza and colleagues, has proven controversial on both technical and ethical grounds. The method comes from the sub-discipline of environmental economics, which attempts to improve, for use in environmental management, the tools that classical market economics provides. The alternative, and to some extent, opposing sub-discipline of ecological economics takes a broader, ecologically-based, view. Whereas environmental economics emphasises efficiency (of resource use), ecological economics emphasises sustainability (of natural capital) and social considerations (Bergh, 2001). We draw on both perspectives in this book, tending to label the first as 'economics' and the second as 'ecology' or 'socio-ecology'. The data for the effects of hurricane Katrina on New Orleans was taken from http://www.msnbc.msn.com/id/9329293 and http://quickfacts.census.gov/qfd/states/22/2255000.html. City population was given as 484,674 in 2000 and 223,388 in 2006.

[Sustainability] Population data for the last 10,000 years is at http://www.census.gov/ipc/www/worldhis.html. The projection to 2040 is from Anon (2004) and assumes that fertility rates continue to fall, so that births and deaths come into balance and the population no longer rises. The energy use figures were taken from Common & Stagl (2005). The figure of 20 Human Energy Equivalents (HEE) per person is a rounded global average, about which there is great variation. For the reader who prefers to think of servile horses rather than enslaved humans, 19 slaves provide about two standard horsepower. The ecological footprint is a measure of relative human demand on the Earth's ecosystems. The calculation starts by estimating the amount of biologically productive land and sea that are needed to regenerate the resources a given human population consumes and to absorb and render harmless the corresponding waste. This is then divided by available land (and sea) area. For 2006, humanity's total ecological footprint was estimated at 1.4 times that of the entire surface area (land, sea and ice-caps) of the planet Earth (Hails *et al.*, 2008) – which is to say that humanity uses ecological services 1.4 times faster than Earth can regenerate them. For example, many, if not most, fish stocks, representing part of **natural capital**, are being exploited faster than they can be replenished (FAO, 2009; Thurstan *et al.*, 2010). Issues for salmon farming in Scotland were reviewed by Anon (2002).

[Humans and the Environment] As discussed in an earlier note, our triad of objectives could also be ascribed to environmental economics (efficiency) and ecological economics (sustainability and equity).

[Governance (*Homo sapiens societatis*)] The subspecies names used here and in earlier sections are whimsical and not part of formal biological nomenclature. However, it is often the case that a proper subspecies name refers to a typical habitat, and this determines the form of the Latin-derived word: in this case, *societatis,* meaning 'of (or in) society' in the same way that *littoralis* means 'of the shore'. The Latin word *socialis* means 'sociable', which appears to be a general attribute of species of the genus *Homo.* Our argument is that, for the last 70 or more millennia, populations of *H. sapiens* have built up social capital in the form of organisations and institutions that were initially vital for survival in a harsh environment. The consequent, human-engineered, environment has become the main shaper of human life and thought. Amongst the developments were technologies (e.g. agriculture) that allowed the more effective exploitation of ecosystem services, leading to further population growth and the appearance and continued evolution of new institutions, including those of governance. Some views of history (e.g., Marx, 1977) see a strict determinism of social institutions by the 'means of production' (and thus by key ecosystem services, as we would see it); others allow for more contingency (Popper, 2002), or, as we would see it, non-linear interactions in the feedback loops between society and ecosystems. Jürgen Habermas points out that in the modern period (the last few centuries) a crucial tension has been between liberal-democratic views, which privilege individual autonomy and underpin 'free' market economies, and civic-republicanism, which gives more regard to the collective (Finlayson, 2005). In essence, as many political philosophers have concluded, individuals are shaped by society, need it to survive, yet can be oppressed by it. Although equity is one of our aims, what we deal with in this book are not radical suggestions for making a greener and fairer society, but what Karl Popper calls 'piecemeal social engineering'.

[Stakeholders] Although we write about stakeholders, the term is better understood as referring to a relationship between a person and an issue – i.e. about stakeholding. Chapter 5 examines Habermas's theory of **communicative rationality** (Habermas, 1981; Finlayson, 2005). Habermas distinguishes the everyday social rules of the **life-world** from the instrumental rationality of what he calls the system of large-scale institutions and organisations. Note that we use the word 'system' with a different meaning, as addressed in the next section.

[Systems] The version of systems theory used in this book combines General Systems Theory, GST (Von Bertalanffy, 1968), ecosystem theory (Odum, 1959), and Soft Systems Methodology, SSM, (Checkland & Scholes, 1990). GST and ecosystem theory understands systems as entities existing in the real world, SSM as mental constructs used for human understanding. Jared Diamond's account (2005) of societal collapse relates it to environmental degradation, climate change, societal adaptability, and relationships with neighbours. It has been critiqued in McAnany & Yoffee, eds, (2009).

[A Systems Approach to a sustainable coastal zone] Where we write 'actors', we could, instead, write 'rôles', which makes two things clearer. First, that what we have described are parts in a set of social interactions, governed by rôle-specific rules (e.g. about how a scientist should behave). Second, that an individual can play more than one part – can, for example, be a government officer during her working day and a stakeholder during the evening. However, an *actor* is a person with motivation and free will and capacity to act in the world, whereas a *rôle* is a set of expectations, and corresponds to a functional way of viewing systems. What we call a 'problem' can also be seen as a dysfunction, which is to say, a deviation from an optimal state in parts of the social-ecological system, judged against objectives of sustainability, efficiency and equity. It could also be seen in relation to the properties of a healthy social-ecological system that includes resilience to externally-imposed change.

CHAPTER 2

Coastal zone problems and the policy context

Audun Sandberg

The coast: ecosystem challenges and policy problems

When you approach the coast, you enter a border zone. Most border zones have a distinct character, as at the edge of a forest, the foot of a hill or the outskirts of a city. But this border zone is different and fundamentally more dramatic than most other border zones. Here three of the classic Aristotelian elements meet each other: the air, the water and the earth (Poole, 2010). The qualitative differences between these elements makes the border-line itself very sharp, although it shifts between high and low tide.

Life in border zones has its problems, but is rewarding at the same time. Most organisms tend to be confined to one side of the border. Terrestrial animals have their feet on the ground and their head in the air, except for the occasional swim. Marine organisms mostly live their lives under water and never visit the dry environment voluntarily. However, there are some crossovers – or connecting species. Reptiles and insects are the most versatile masters of both water and earth. Marine mammals, also, utilise all three elements to varying degrees, getting all their food from the wet environment, breathing the air and, except for whales, breeding on land. Some seabirds can live almost all their lives at sea, and get all their food from the sea, but they depend on the air for transport from one coast to another and on the earth for nesting. In the zone of alternating wet and dry between high and low tide, there are myriads of specialists who favour this periodic and dynamic environment, and are a source of food for both terrestrial and aquatic organisms. These seashore creatures are probably the most typical coastal dwellers of them all (EU Commission, 1999).

This brief glance at the coastal web reveals a breathtaking complexity. Every little creature is connected to some other creature in some way or another, directly, or indirectly through long lines of interactions. As shall be seen, these interactions not only take place within the confines of one element alone, but cross the seemingly sharp border-lines between water, air and earth. And their effects can extend across the blunter border-lines between ecological systems and social systems (including their economic, legal and political aspects). The ecological interactions can run smoothly and seemingly balance each other so as to produce harmonious and pre-dictable coastal ecosystems that support human coastal communities in a sustain-able way. Or they can run wild, exploding into algal blooms, or generating invasions of sea urchins and crabs that occupy the ecological 'space' of important commercial cod stocks (Kurlansky, 1998). Or they can 'run dry' and implode into mysterious collapses in crucial capelin stocks or into failures in the supply of small fish that feed valued seabirds like the puffin (Figure 2.1) during their breeding season (Anker-Nilssen *et al.*, 2007).

When we try to build a model of this immensely complex border zone and create a representation of this as a coherent and integrated functional system, we encoun-ter a number of problems. Some of these are of an epistemological nature: we do not yet have sufficient theory and knowledge to be able to describe, fully, the behaviour of the whole coastal ecosystem – with its tentacles extending into the three different elements. But as we shall see in this book, advances in the eco-sciences and in the

2.1 Puffins on the Island of Hornøya, Norway. Non-human users of ecosystem services (fish) and used in this book as symbolic of natural systems.

understanding of ecosystem dynamics have provided a common basis of knowledge that can be applied across the border zone.

But some coastal zone problems are of a different character. They arise from human affairs, or, more precisely, from the poor rules and bad policies with which humans try to regulate their use of ecosystem services, and which are not necessarily helped by the advances in coastal ecosystem knowledge. As used in this book, **policy** means 'a deliberate plan of action to guide decisions and achieve rational outcome(s)'. Plans have, of course, to be made and implemented by people, but, as we shall see, people who are often grouped into organisations and who act under the aegis of institutions. Collective tools or social functions can be called 'instruments'. Some of the problems in developing and applying policies arise because policy instruments, like Governance, Management, and Planning, tend be confined by not only the physical borders between land and sea, but also the socially constructed borders between different sectors of human activities: Fisheries, Aquaculture, Agriculture, Transport, Industry, Defence, Harbour, Recreation, etc. The kind of problems we are now considering, those in the human world, are often 'owned' by different government ministries, have their own sectoral legislation, their own brand of expertise, their own organised interest groups, and their own preferred solutions. Depending on the sectoral perceptions, some of these coastal problems are considered as real and existing in the physical environment, while some are deemed to be virtual problems and hence capable of correction by changing people's understanding of the issues. The distinction between these can often be a tricky political matter. But as we shall also see in this book, recent advances in the sciences of governing societies make transparency of problems and the integration of ecological science-based knowledge and policy-related knowledge more feasible than ever before (Poteete *et al.*, 2010).

The occasional visitor to a coastal resort will usually not notice the dynamic character of the coastal systems. The pattern of coastal activities will seem stable and predictable to him or her. Coastal fishing boats leave their harbour and go to sea at the break of dawn. Later, schoolchildren go to school and terrestrial workers enter offices and shops, while beach lovers slowly move to the edge of the water. After lunch the first fishing vessels arrive with the catch and the receiving station or the auction hall comes to life. After that the schoolchildren and workers return home for dinner – and another coastal day is over. The visitor who stays for a few days might begin to see that the movement of the boats varies slightly according to when high and low tides occur. If he stays even longer, he might observe that both the quantity and the composition of the fish catch varies with the season and the migrations of important fish stocks like herring, cod, mackerel and tuna. If he is invited to spend some days on board one of the fishing boats, he will witness a flow of information from the ecosystem to the little human society on the boat. Some is transmitted by way of the sonar and the fish finder, some comes from the catch itself. He will hear much similar information from other boats on the marine radio network. Have herring been observed, where and at what depth? What is its quality, i.e. what has it been feeding on and consequently what price would it fetch in the market? The fishing master will process and interpret all this information, mix it with his own experience and thus plan the boat's fishing strategy for the coming days and weeks (Barth, 1966).

Thus we understand that the coastal system is not only more dynamic than it appears at first glance, but that this coastal village is nested in larger ecosystems of migrating fish and ocean currents, and is immediately subject to the effects that eco-system change has on supply in larger social systems of fish markets where demand also fluctuates because of changing tastes. Less easy to glimpse, and harder to understand, are the connections between the village, the market, and the slow political processes of setting catch quotas and changing investment conditions.

One class of coastal problems with policy relevance can thus be connected to this fluid and dynamic character of the coastal environment and the challenges this poses to coastal communities and coastal regions. Conveniently we can divide these into the short term (daily) tidal action, the medium term (yearly) seasonal variations in the abundance of marine organisms, including shellfish and migrating fish, as well as migrating seabirds and tourists, and the long-term variations of growth and decline in important commercial fish stocks (20–100 year perspective).

Another class of coastal problems is connected to the influence of social and economic dynamics on crucial coastal ecosystems. Often these problems result from the aggregate effects of small individual and seemingly innocent human actions. Agricultural runoff and sewage leakage from holiday homes might each seem negligible, but the overall effect on the water quality and the frequency of harmful algal blooms in the Baltic Sea might be dramatic (Elmgren, 2001). The gradual destruction of coastal reefs by fishing gear and the consequent loss of habitat for keystone species is another example (Jennings & Polunin, 1996). Over-fishing and over-harvesting of marine resources also has this basic character of collective irrationality – leading to increasing **transformation costs** and **transaction costs** (North, 1990) for social systems, but often resulting from small actions that, on the individual level, seem rational.

It is in a series of attempts to achieve a higher degree of collective rationality that most coastal nations during the last 20 years have introduced catch quota systems for wild fish in one form or another. Despite concerns about the wasteful discarding of 'by-catches', these usually work for resource management purposes, but have several unintended consequences on coastal social systems, e.g. closure of entry to coastal fisheries and marginalisation of coastal youth (Maurstad, 2004).

Other coastal activities have similar problems. Fish-farming, including the cultivation of mussels, oysters, salmon, sea trout, bream, halibut and cod, depends highly on the quality of the coastal marine ecosystem. At the same time, open cage aquaculture imposes considerable stress on these ecosystems. It is both the feed waste and the faeces from the concentrated biomass in fin-fish farms that endangers crucial ecosystem services, and then through complex ecosystem interactions becomes a threat to the surrounding wild marine biota and in turn to the fish farm itself – and to neighbouring aquaculture enterprises (Tett, 2008). If there are no institutional arrangements to curb the individual temptation to expand the aquaculture enterprise in times with high prices, social science theory – and experience – tells us that coastal aquaculturalists can be their own worst enemies.

The opposite effect can also be a coastal problem. In the Danish Limfjord the increased nutrient load has replaced the traditional fin-fish fisheries with a very profitable mussel fishery. With the implementation of the EU Water Framework Directive, the Limfjord has to be 'cleaned' – with dramatic decrease in the productivity of wild mussels as a result. Here a threat to the adaptation of social and economic activities to a humanly modified ecosystem becomes a larger-scale coastal problem because European Union and Danish environmental objectives are in conflict with local concerns for livelihood (Dinesen et al., 2011).

Wild fish and farmed fish also interact through other pathways in social-ecological systems. In the North Atlantic, the dramatic recovery of the wild cod stock has lowered the prices for cod, the baccalao and the stoccafisso to a level where farmed cod can only barely be produced profitably. For the cod farmer who optimistically invested when the wild cod stock was at a low ebb and now has to give up the business, the healthy state of the wild cod is therefore as much a coastal 'problem' as the mutual pollution from the fish farms.

What we call 'coastal problems' therefore contain all kinds of different problem categories. Some of these problems are successfully communicated, into the human systems of social discourse and policy formation, by information about pollutant concentrations in relation to legally-defined threshold values, or about changes in the biological community, as specified in the EU Water Framework Directive. Some problems are caused by strife among human agents alone and do not involve any need for **Integrated Coastal Zone Management (ICZM)**, just bold political decisions. Some coastal problems arise in the ecological systems but are not communicated to the social system before it is too late to remedy them. Famous systems theorists have argued that this is because these different systems have too few 'connectors' that can transmit information about poorly functioning – or dysfunctional – social-ecological systems interaction (Luhmann, 1989). Resource users are often muted by strategic protection of vested interests, and environmentalists are often marginalised by national power structures. But in the policy-making rooms, it is the volume (loudness) of this communication that tends to be the decisive factor, not the fundamental character of the problem.

Travelling with a salmon

A journey with wild salmon through the different worlds of the coast will show in a simple way how the complexity of the social-ecological system works on the coasts of contemporary Europe. We know that all salmon farmed in Europe originates from the genes of the wild Atlantic salmon. As an anadromous fish it has been selected for a difficult way to live and reproduce; feeding in the North Atlantic and spawning in a distant home river, often with considerable risks for a low reproductive success compared to, for instance, the lazy cod. Evolution made it one of the strongest fish on the planet, and thus it became the anglers' favourite and its muscles a favourite food. More than 100 years ago enthusiastic anglers developed a way to hatch and breed salmon to enhance the natural salmon stocks in rivers where the suitable salmon spawning areas were dwindling, due to hydropower development, flood control works, agricultural expansion and riverside road-building. In spite of scepticism from genetic 'purists' insisting on keeping river-specific brood stock apart, these stock enhancement programmes were quite successful and kept the wild North Atlantic Salmon stock at high levels through most of the twentieth century. Why then is the wild Atlantic salmon a threatened species today?

The travelling salmon connects in a nice way the oceans and the rivers through the coastal system and can thus serve as an indicator organism of the health – or the peril – of a particular coast. Its need for clear, well-oxygenated water made the passage of wild salmon a good indicator of the water quality in estuaries. Thus, the slogan 'the salmon is back' has become the sign of success for clean-ups of north-European river systems like Mälaren in Stockholm and the Rhine in Germany/Netherlands.

On its way from the ocean, the well-travelled salmon first meet the ocean fishers, and, although drift-net fishing for salmon is now banned in all European countries except Ireland, a large amount of salmon is still caught offshore. When entering the archipelagos and fjord systems of the formerly glaciated coasts of northern Europe, the salmon more and more often meets the cages of farmed salmon, emitting strange pheromones that disturb the efficient homing of the wild salmon. Here are also the large concentrations of sea-lice as a result of the high concentration of salmon in the cages of the fish farms. As a parasite it sticks to the wild salmon and can weaken the strong fish so that it does not manage to combat the crucial waterfalls and reach the best spawning sites.

In the salmon rivers, the anglers are eagerly fishing for salmon, but with the spread of 'catch and release rules', the effect on the stock is dampened. On the spawning site, the spawning itself is often disturbed by runaway farmed salmon that clumsily interfere in the mating ritual and excavate the carefully covered eggs. As hydro-electric power stations are run with different and more profit-motivated regimes, the annual spring flood from the snow melting in the mountains of Scandinavia and Central Europe is often curbed in many rivers and the natural hatching of the eggs is increasingly at risk. When the young salmon are ready to travel towards the ocean, they face a number of non-traditional hazards. One of these, currently causing much concern, is the zone of sea-lice from farmed salmon that awaits them when they reach the coastal zone. The attacks by these parasites have been known for a considerable time to be fatal for young salmon (Grimnes & Jakobsen, 1996). Only a fraction of the eggs from the spawning thus grow up into young salmon in their distant oceanic feeding grounds.

From the perspective of the wild salmon these travels take them through a number of different institutional worlds, with vastly different property right regimes, vastly different rules for catch and emissions to water, and policy responsibility fragmented between many different government agencies – often belonging to several countries. Efforts in some countries to create special protected salmon runs and 'national wild salmon fjords' without aquaculture (Government of Norway, 2006), have met heavy resistance from the powerful multinational salmon farming industry.

Thus, this journey with the wild salmon illustrates the difficulties of making coherent rules that benefit one of the most typical elements of the larger coastal ecosystem. The wild salmon's complicated life journey ties together the sea, the coast and the rivers inland. Thus, many of the problems that face it arise from the fragmented nature of human institutions, and can only be solved by the social system.

Representations of coastal zone problems

To get a handle on complex problems, you can take a piece of paper and make a drawing of all the elements you think make up the problem. Then you can go one step further and draw lines between the different elements. You can use thick lines where you think the relationship between the problem elements are strong, thinner lines where you think they are weaker. This is a simple representation of how you perceive a coastal problem to be constituted. As the coast is complex, you will, for your representation, probably need a large sheet of paper. We will later show how this can be developed into a conceptual model, and how this sketch can then be formalised and elaborated into a model system (a virtual machine) and put to practical use in a scientific or policy-making context. This is the main theme of the following chapters in this book, especially chapters 3 and 4. Usually you will have to decompose the complexity into manageable components, but in such a way that the multiple interactions are not lost or reduced to trivial nonsense (Ostrom, 2007).

But before starting an analysis of this sort, it is of crucial importance to be aware of the nature of a problem. Coastal reality is usually highly complex. In order to be able to refine this complexity, we must be able to judge whether this is the kind of coastal problem that can be modelled – or not – and to what extent the problem description can be refined without losing the essence of its complex nature. Although, as shown in the examples above, coastal problems can be of widely different character, it turns out that all can be fitted into one of three classes:

❑ The first class is that of cases in which humans inflict damage on coastal ecosystems, either collectively by bad policies or individually by uncoordinated actions, based on selfish rationality that, taken separately, appear as innocent. The chain of cause and effect from the human activity to the damaging consequences can be long, and the aggregate long term damage can be grave and irreversible. But the cause of the problem is here to be found in the way the social or economic system is organised, and in the institutions that constitute the rules for conduct in these social organisations. A well-functioning social system has feedback mechanisms that can correct such rules before threatening environmental damage occurs. If such self-correcting mechanisms are lacking, the social system is functioning poorly or is non-functioning. In extreme cases a social system is 'producing its own poison', i.e. developing towards its own breakdown. It is then dysfunctional or malfunctioning as a social system (Merton, 1968). Deterioration of the coastal ecological systems might be one of several explanations for this, through the pressure a dwindling resource base imposes on the individual and corporate actors in the coastal social system.

❑ The second class is that of problems in which nature causes difficulties for coastal communities, e.g. by changed composition of species, algal blooms, disappearance of fish stocks, etc. Although the cause of the problem lies in a part of the ecosystem that is outside human control, nature is still innocent. It is the actions of individuals, collectives and political representatives that have made the communities and the social systems vulnerable to these kinds of ecosystem processes, by eroding the social system's resilience to disturbance or capacity to adapt to change. In contrast to those problems where the coastal ecosystem can be brought back to its healthy state, i.e. by rebuilding a fish-stock (as above), the integration of science into policy is more difficult when the task is to reduce vulnerability, strengthen adaptive capacity and build robustness in coastal communities. Often, new sets of institutions have to be designed to cope with a breakdown in traditional coastal resource use. The challenges of finding sustainable solutions to such complex problems are discussed in detail in chapter 5 in this book.

❑ The third class is that of coastal problems that do not involve any interaction between ecological and social mechanisms. Where one group of coastal users, for instance ocean-going trawlers, fish offshore on migrating fish that are on their way to a site inshore where coastal fishers are waiting, there is a conflict between two parts of the social system. Such upstream/downstream conflicts do not require any advanced modelling or analysis of interaction of social systems with the coastal ecosystems. Problem solving merely requires bold policy-making that establishes rules and regulations that can distribute the total allowable catch in a just and legitimate way.

Before we depart from our open and fuzzy conceptualisation of the coastal world, we are still able to take on board new explanatory factors and incorporate them in an increasingly complex representation of the coastal systems. As a large number of the typical coastal problems are 'crossover problems' that originate in one social, economic or ecological sector and spill over into other sectors, the open character of this initial process is very valuable. It facilitates curiosity-driven conceptual testing of possible relationships before the process of description goes into a more closed phase of formalisation. It is in this initial phase that it is possible to detect the underlying logic of the mechanisms at work in a particular coastal setting. Such underlying logics can often be represented diagrammatically, as in Figure 3.4. If their schematic structure is recognised by both scientists and stakeholders as a 'reasonable' representation of the problem and of the dynamics of the interaction of the coastal systems, this can be a vital building block when we attempt to refine the representation and illuminate the strongest and most decisive factors at work. Such underlying structures can have the character of a simple functional logic with a negative feedback that contributes to a self-correcting system (Stinchcombe, 1968). Or they can have a character of frequent positive feedback loops that make the system explode or implode – or even reach a tipping point, where the system changes into a qualitatively different system. Or coastal problems can be represented by the antithesis of hierarchical order, a panarchical order of interlinked and continual adaptive cycles of growth, accumulation, restructuring, and renewal (Gunderson & Holling, 2002). These underlying logical structures of problem representations and explanations are further exposed in the next chapter.

In coastal sites around Europe, we can see how different coastal problems can be represented as different classes of problems, and also how these problems can be seen as having different logical structures. In the case of Limfjorden in Denmark, the problem goes beyond a contemporary conflict issue between mussel fishers and environmental concerns. The Danish regulatory authorities are trying to comply with the EU Water Framework Directive and clean the Limfjord of agricultural run-offs. This is a coastal problem where human activities (modern agriculture) have been driving ecosystem changes, with the danger of **eutrophication**. But the interaction of the ecosystem and the social system has here surpassed the 'tipping point'. The more nutrient-rich waters have created the productive base for a new resource, mussels, both wild and cultivated. The latter have led to an emerging mussel aquaculture and the former to a profitable mussel fishery – as well as large colonies of mussel-eating water-birds. The social system has thus overcome its initial vulnerability to ecosystem change and is now well adapted to the new state of the fjord's ecosystem. This is the background for social resistance towards legally-required attempts to push the ecosystem back to its perceived natural state before the tipping point (Dinesen et al., 2011).

In the case of the Lagoon of Venice the coastal conflict issue is between unsustainable clam fisheries and potentially sustainable aquaculture, coupled with consumer health risks from marketed clams. As a coastal problem, this can be represented as a conflict between different users of coastal resources and a lack of governing rules for harvesting and quality control in connection with marketing. Strictly speaking, the representation of this coastal problem does not require any analysis of interaction between coastal ecological systems and coastal social systems, except for some form of assessment of the overall 'social carrying capacity' of the Venice lagoon clam resource: i.e., what is the capacity of the lagoon to produce clams, and how many people can be supported by this capacity? As a study during the SPICOSA project showed, when new harvesting rules are introduced and new, properly managed, clam farming systems are in place, the problems of the Venice lagoon tend to solve themselves (Melaku Canu *et al.*, 2011).

A common trait in all these examples is that human agents, willingly or unconsciously, have modified the coastal ecosystem to suit their various economic activities. Very often the consequence of such modification is a simplification of the coastal ecosystems, and thus the modification done by one group is often detrimental to the interests of other groups. The social-ecological systems that have evolved for cultivating the coastal seas, for maricultures and aquacultures, have in many cases improved the conditions for cultured activities, but made them worse for fishers depending on catching wild fish. The increased fertilisation that benefits some mussel farming areas, and the eradication of predators that might otherwise harm fish farming operations, are important cases here. But in other cases the marine cultivations do not only create problems for wild ecosystems, but also for themselves; as in cases where the monoculture of a species at high density constitutes a hazard to the productivity of the cultivation itself. As we saw above, salmon lice are given ample chance to multiply in the densely stocked cage farms and can thus threaten the entire harvest of salmon.

When a coastal problem has these characteristics, it is very difficult to work out policy solutions that can solve the problem. This is in contrast to typical coastal degradation problems, where a coastal enhancement programme, a conservation decision, a ban on a certain harmful use, or a temporary moratorium on the catch of certain species, enables nature to restore itself. In cases where human systems for a long time have interacted with coastal ecosystems and carved out a 'second nature' for themselves (Rousseau, 1754), policy solutions are much harder to find. The reasons here are complex, and contain, in addition to a number of ecosystem interconnections, also important social and economic elements. The investments made in cultivation infrastructure for mussels or fish are not mobile like a fishing vessel. They are, literally, **sunk costs**, often incurred on credit and with collateral in the owners' houses. A change in the rules for the groups depending on these modified coastal ecosystems would therefore, in many cases, spell bankruptcies and social disaster. As stakeholders in a policy-making process, such groups usually have stronger vested interests than others such as environmentalist groups advocating more idealist, or public, interests. Since so much is at stake, they will be prepared to go to great lengths to have policy measures designed in a way that enable them to continue their cultivation operations. However, sometimes ecosystem malfunctions accumulate to such a degree that unpopular policy measures have to be taken and certain operations have to be ruled out as illegal.

In such cases stakeholders will often claim that they have invested in accordance with a certain government policy, maybe even with the protection of a licence and a government subsidy scheme. When the policies then are changed, they will claim to be caught in a social trap, where they have fallen victim to arbitrary policy-making. If such sentiments become widespread in coastal communities, these can have long-term consequences when it is time for governments to be re-elected. Policy-makers therefore tend to be cautious when they start policy processes that might fundamentally change conditions for coastal enterprises and employment. Even when the environmental consequences of continued operations become obvious, these kinds of coastal problems are characterised by a considerable amount of political impotence, i.e. a reluctance to take political action. Only when the environmental misery – or the user conflicts – reach a certain threshold, can a new – and legitimate – policy-making process be initiated.

Policy challenges in coastal zones

Before opening the big black box of Science–Policy integration, it can be useful to ask what policy-making is all about, apart from the obvious objective of solving different kinds of coastal problems as described above. In the first chapter we defined the whole chain of governance as 'steering', setting the course, navigating the dangerous stretches, and reaching a destination. But will it turn out, when the governed get to provide their opinion, that this destination was where they really wanted to be? Policy-making, as a fuzzy concept, is used to label various stages in this social steering process. In former times, lofty discussions on where to set the course, and what were the most worthy objectives of societal development, were often called policy-making. Today, these more ideological debates are taught as 'political philosophy' in universities around the world, and it is now rare to see parliaments having these kinds of debates on a regular basis, where the ideological attractiveness of individual freedom, equality, free enterprise and solidarity are brought into the policy-making process. This is mainly because in our late modern age, the belief in utopian social models is no longer in fashion: we no longer have a collective vision of a perfect society somewhere ahead of us. So contemporary policy-making is not about deciding where the final harbour is, but about navigating between all the crises and temptations of modern societies, between greed and fear, to find the optimal solutions to problems that are mainly of a non-ideological character. Thus the emphasis of modern policy-making is more on how to move ahead rather than where to get, and sustainability tends to be more about 'keeping afloat' than about seeking the final solution in a perfect balance between humans and nature.

The problems of environmental degradation, the problems of climate change, the problem of resource mining, the problems of over-investment in harvesting activities, even the problems of ageing coastal populations, are typical of modern challenges to policy-making. The interlinkages between these kinds of problems are often complex, especially when ecological and social processes interact. This kind of complexity has increased in the last 50 years, mainly because the wealth of knowledge has increased and is today the main challenge for policy-making as a navigational pursuit. How to deal with this complexity is thus a major test for modern policy-makers – in a policy-making environment that seemingly is constituted as a string of single, simplified and, to a large extent, unconnected decisions. As an answer to this challenge of handling complexity, we find it useful here to regard the essence of policy-making related to coastal areas as 'rule-making' related to the human use of coastal resources. Rules, whether formal or informal, will always come into existence where people interact over the use of some resources. This is human nature, but the results can be complicated. When social scientists map resource-sharing rules in a coastal community or region, they find that they can be prohibitive, prescriptive, or have the properties of an incentive. The rules can prescribe what individuals may do on their own and when they must come together and act as a collective. The rules might have organised monitoring and sanctioning built in, or might lack such mechanisms.

We may, in optimistic moments, call this kind of policy-making activity 'crafting institutions for sustainable development'. All such rules have to be considered as temporary, because the quality of policy-making as rule-making is not known until it has been tested in real life conditions (Ostrom, 2005). Policy-making therefore has an experimental character: the best possible combination of rules for handling complex coastal situations should be thought out using the available

41

scientific knowledge and scientific models. If the rules fail, the lesson learned is not to repeat those combinations, at least not within the same generation. But despite the accumulation of experience, which implies that rules should get better and better, most policies are good only for the time being, because they apply to dynamic and ever-changing social and ecological systems. Contemporary policy-making thus consists of the experiments that human societies do on themselves.

With increased complexity, the science content of policy-making tends to increase and the quality of the science input into policy-making consequently comes under more open scrutiny. This means not only that the methods used and the data gathered must be of a recognised quality, but also that the research questions used for framing the hypothesis, the underlying assumptions, and the parameters chosen, should be illuminated and shown to a public who are increasingly well educated and critical. The worst nightmare for an ambitious policy-maker is to be confronted with high quality scientific knowledge that conflicts with that on which she has based her decisions. This is often the basic difference between a national government and an opposition; the policy-makers in power have usually better access to the research centres with the best reputation, and thus the scientific base for their policy solutions tend to appear as more trustworthy. On regional and local levels of governance the balance might be reversed, as the local authorities often have to use consultants to provide the basis for their policy-making, whereas the national environmental protection organisations that they meet in the coastal zone often have better access to prestigious research laboratories.

A criterion that is often used for evaluating the credibility of scientific contributions to policy-making is the notion of truth and objectivity. An objective scientific finding from a renowned research laboratory is believed to carry more weight than the personal opinion from a lone scientist. Thus the policy-making process is full of safeguards and defence mechanisms to protect its outcomes against criticism of the basis of decisions. Here the objectivity – or at least the consensual subjectivity – of the research, as well as the reputation and prestige of the researchers, plays an equal role. Science itself has institutionalised a number of mechanisms to guarantee the quality of its input in the policy-making process, including 'blind' peer review processes, scientific panels, and research committees, which make some scientific results appear more weighty than others.

Without entering the centuries-long philosophy-of-science debate about the difficulties of obtaining objective knowledge (Popper, 2002), we can safely assume that scientific knowledge in some way or another has to be given a quality stamp or some form of authority guarantee in order for it to be usable as a basis for policy-making.

On the one hand, there are good examples of sciences that go to great lengths to establish such guarantee mechanisms in order to persuade governments and the public that their knowledge is close to factual truth. For example, the Inter-governmental Panel on Climate Change (**IPCC**) is the leading international body for the assessment of research on climate change. Through its network thousands of climate change researchers strive to establish facts about the effects of current climate changes, and its organisation is tailored to guarantee scientific authority:

> 'Because of its scientific and intergovernmental nature, the IPCC embodies a unique opportunity to provide rigorous and balanced scientific information to decision-makers. By endorsing the IPCC reports, governments acknowledge the authority of their scientific content. The work of the organization is therefore policy-relevant and yet policy-neutral, never policy-prescriptive.' (IPCC, 2010).

On the other hand there are some sciences that openly claim that objectivity is impossible at the outset, and thus risk being acceptable only for certain kinds of policy-making. For instance, postmodern standpoint theory claims that:

❑ A standpoint is a place from which human beings view the world;

❑ A standpoint influences how the people who adopt it socially construct the world;

❑ Social group membership affects people's standpoints;

❑ The inequalities of different social groups create differences in their standpoints;

❑ All standpoints are partial; so for example feminism as a point of view can coexist with other standpoints (Smith, 1999), and ecofeminism (McGuire & McGuire, 2004) differs in its theory and practical outcomes from a male-dominated paradigm of ecology.

When sciences with such different outlooks on the nature of scientific know-ledge are to work together to provide the basis for policy solutions to the com-pounded problems of coastal zones, the challenges might seem almost as great as when trying to understand the intricate connections between ecological systems and social systems. This cleavage between the different strands of science has been termed the *two cultures* and has entered the legacy of science as a shorthand for differences between two attitudes. These are, on the one hand, the increasingly con-structivist world view suffusing the humanities, in which the scientific method is seen as embedded within language and culture, and on the other hand, the scientific viewpoint, in which it is believed that the observer can objectively make unbiased and non-culturally embedded observations about nature (Snow, 1993). When C.P. Snow coined this distinction in 1959, he was himself not totally unbiased; he was arguing for a greater role in policy-making for the natural sciences, at the expense of the literary intellectuals (humanists) in the modern world. Since then the rift has not diminished, and between the disciplinary university faculties we still find mutual incomprehension tinged with hostility.

But in the practical research work for understanding and solving the new and complex challenges in the field of sustainable development, scientists from widely different disciplines have to come together, each with their funding paradigms and their own methodological toolboxes. The fundamental works of Thomas Kuhn and Karl Popper have provided all sciences, natural and social alike, with tools that facilitate such cooperation. Mutual understanding of, and tolerance for, competing and shifting paradigms within disciplines can contribute to build trust among scientists from widely different disciplines (Kuhn, 1996). And an increased awareness of the necessity of an explicit preconception of what is to be observed can contribute to set a complex research project on the right track at an early stage. (Popper, 2002). In systems thinking, the long-standing *General Systems Theory* (von Bertalanffy, 1968) and the hard science that goes along with it, is being complemented by soft system methods that allow for multiple conceptualisations of the world. (Checkland & Scholes, 1990).

Because not all paradigms in one discipline (e.g. ecology) are compatible with all paradigms in another discipline (e.g. political science), different matches between the disciplines can give rise to multiple perceptions and interpretations of social-ecological systems, which can lead to different – and innovative – conceptual models. Contrary to what was possible during the eighteenth-century Enlightenment, no single scientist can any longer be an expert in multiple disciplines, nor can she be a master of all multiple methods used within a single discipline. Thus more and more practical research work is done by research teams composed of scientists from different disciplines who can complement and stimulate each other (Poteete *et al.*, 2010). Such interdisciplinary and multi-methodological involvement is demanding for all scientists and requires a mutual openness to the other collaborating scientists, as well as confidence in their own field of expertise. As more and more environmental problems become recognised as multisectoral and crossover issues, the future development of most sciences will be strongly influenced by such practical research collaboration, both in the framing of crucial research questions and in the interpretation of research findings. With more transparency in the research processes, both within research teams and between these teams and coastal communities, the objectivity of the natural sciences will also, to an increasing extent, be questioned by the policy-makers and the stakeholder groups that are at the receiving end of the science–policy integration. The scientists' standpoints in environmental issues like conservation, biodiversity, eco-efficiency, climate change and genetic manipulation will also, to an increasing degree, have to be made explicit for such interaction to be seen as legitimate.

When making true-to-life representations of coastal problems, it is tempting to include also the effects of stakeholder actions and policy-making decisions in the modelling exercises. Any individual or collective action aimed at solving a coastal problem, any change of rules, has consequences that affect the interplay between ecosystems and social systems. To some extent these effects can be studied by modern social science methods and modelled through game-theoretical models or through agent-based modelling, which can then be attached to models of the ecosystem functioning (Poteete *et al.*, 2010).

However, to represent all the possible ingenious human actions – and all the systemic outcomes of these – in a formal model, is a formidable task that should be undertaken only for very small and well-defined coastal problems. As resource users, humans are fairly predictable: without rules they will usually go on exploiting the resource until it is empty or marginally unproductive (Hardin, 1968). With good 'rules-in-use', humans will tend to abide by these and be able to succeed with a sustainable resource use for long periods. The tools available for making such good rules under diverse social and ecological conditions have been greatly improved during the last decades under the name of Institutional Analysis (Ostrom, 2005). Better tools for analysing complex social-ecological systems (SESs) are also being developed. In these analyses an SES is seen as made up from Resource Systems with Resource Units, which interact with Governance Systems and Users or Actors to produce Outcomes that in turn are evaluated by the Users/Actors (Ostrom, 2007).

In the field of policy-making and the crafting of institutions, *Homo sapiens* is the most inventive of all species on Earth. But humans can surprise themselves both in their individual and collective actions and in the outcomes of these actions. Thus it is as important to try to model the effects of policy on human uses of coastal resources, as it is to model the ecosystem changes. In neither sort of modelling does simple logic work well; because of the generic system properties that we will discuss in chapter 3, both ecosystems and social systems can change either less or more than predicted on the expectation of a linear response to disturbance. Quite often, social systems do not show any reaction to a gradual degradation of the coastal environment, until the latter has reached a certain painful threshold (Figure 2.2). After falling over this threshold, the human reaction is often massive, both in political and institutional terms. Sometimes it is an overreaction. Conversely, human anticipation, social fear or activist lobbying can often start policy-making processes before an environmental problem has shown any significant effects in a coastal system.

These complexities have the result that formal representations of human actions in the policy field have rather low predictive capacity and that they therefore are

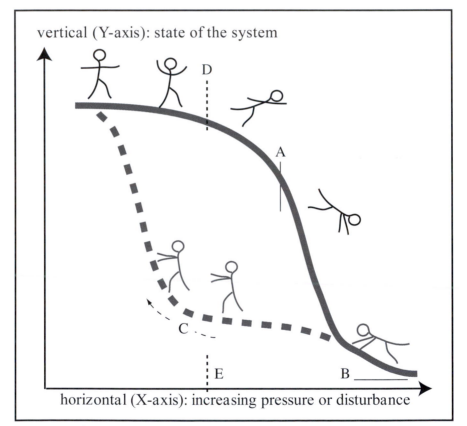

2.2 Cartoon of ecosystem collapse. Systems are supposed to be stable against a certain amount of disturbance, but when perturbed beyond a certain point (shown by A in the cartoon) they collapse into a different state (B) that is degraded compared with the starting point and provides fewer services for humans. Theory (Krebs, 1988; Scheffer et al., 2001) suggests that this alternative state may itself be stable, so that a large relaxation of the pressure or disturbance (to E) might be needed before recovery can occur (C). System managers would be prudent not to allow system state to go beyond D, because of uncertainty about the exact position of the edge of the cliff.

difficult to include in comprehensive social-ecological models of coastal systems. One obvious solution is therefore to keep the 'black box' of policy-making outside models of coastal social-ecological systems, but arrange to have several consultation points along the way. At the first of these way-points, the politically most pressing coastal problem can be identified by a process called **policy-issue mapping**. At this stage it is also possible to identify the main groups in civil society that have a stake in this problem, and to map the degree of their vulnerability, and the intensity of their engagement, with **policy-stakeholder mapping** (Vanderlinden, 2011). At the next way-point, stakeholders and public officials can meet to deliberate on and agree possible solutions at either the operational or collective levels of governance. These solutions can, as we have seen, sometimes be simple – for example, building a more effective plant for the treatment of sewage – but sometimes it will be better to see them as comprising alternative futures involving complex scenarios, as will be discussed in the next chapter. Finally, when the systemic interconnections have been established and a dynamic generation of possible futures has been made using simulation models, stakeholders and policy-makers can get involved in deliberating and making choices regarding politically favoured futures. This final way-point is dealt with in detail in chapter 5.

If the model representing the social-ecological interactions is good enough, the regulatory and institutional tools necessary for achieving this preferred future should be clearly identifiable in the model. In the case of nitrogen discharges or runoff from farms and households to Himmelfjärden in eastern Sweden, solutions to complex problems were found through this kind of iteration between model representation of the coastal problem and frequently repeated consultations with involved stakeholders (Franzén *et al.*, 2011)

In most real coastal policy-making situations, the use of mapping and deliberation tools will in most cases be tied into existing governance tools. These include: coastal area, or maritime area, development planning process; and regulations requiring permissions to be given for harvesting of a certain coastal resource, or for cultivations in a certain coastal ecosystem.

Thus it is, in general, the local and regional coastal elected bodies and policy-makers that have the practical task of reconciling the variety of ecosystem considerations and human interests in their coastal areas. It is therefore for them that most of the modelling tools and the policy consultation tools explained more in detail in chapters 4 and 5 are of the highest relevance. Policy-makers at the national level are often sectoral policy-makers who have little contact with the compounded problems of a typical coastal location. At the level of the European Union, the attention is more towards the procedural questions related to the sustainable governance of the European coast: how to get institutions for sound and effective integrated coastal zone management into place by means such as the Water Framework Directive, the Marine Strategy Framework Directive (and others listed in Table 1.1), the EU Coastal Policy decision, the Union's Maritime Spatial Planning Policy, and so on.

Identifying coastal problems

We have now seen something of the variety of problems that may afflict social-ecological systems in the coastal zone and something of the variety of people who are impacted by them or have to deal with them. We will now look further into a number of useful tools, methods, or procedures that can be employed for focusing and clarifying coastal problems, so that the work of finding solutions may begin. There are three categories of such tools:

❑ DPSIR, or similar, analysis of driving forces working at the system level;
❑ Stakeholder- and institutional-mapping;
❑ Deliberation support tools.

DPSIR has been described by Luiten (1999) as 'a chain of linkages between the driving forces within society (D), the pressure on the environment (P), the state of the environment itself (S), the impact on people and nature (I) and the desirable response (R)'. The SPICOSA project worked with the similar idea of 'cause-and-effect' chains, leading from a *human activity* to an *impact* on the supply of ecosystem services to humans. For example, the human activity of aquaculture might lead by way of the problems of eutrophication that can result from nutrient overload in water, problems including a loss of certain ecosystem services, such as the water clarity that is highly valued by Scandinavian people. In the case of nitrogen discharges from the Himmer fjord's sewage treatment plant the 'driver' is the human activity of living, which generates nutrient-rich waste that is flushed into wastewater treatment systems that in turn discharge some of it into the sea. One management option here is to re-convert low-lying agricultural fields back into wetlands in which soil bacteria remove nutrient-nitrogen by converting it to nitrogen gas. This is a strategy that went beyond a simple DPSIR analysis (which might merely point to a more stringent level of sewage treatment) and became the outcome of a process involving identification both of stakeholders and of perceptions of the problem (Franzén *et al.*, 2011).

In most coastal regions, and depending on the type of problem, there are many different categories of stakeholders with legal rights and moral interests of various strengths. And there are policy-makers with various degrees of influence and power. The next step in analysis of a coastal zone problem is, therefore, to investigate who are involved as stakeholders and policy-makers, using the tools of stakeholder mapping, institutional mapping, and issue mapping.

Institutions are systems of rules and procedures, both formal and informal, that structure social interaction by constraining and enabling actors' behaviour. The 'actors' mentioned here are people doing things, such as (in our context), harvesting clams in the Lagoon of Venice (Melaku Canu *et al.*, 2011), reviewing proposals for new sewage treatment schemes in Barcelona, or putting on their scientists' white coats as they enter a chemical laboratory to analyse water samples in Izmit (Tolun *et al.*, 2011).

It is a common feature of human societies that institutionalised ways of doing things give rise to organisations that sometimes seem to have 'a life of their own', perpetuating themselves long after the departure of those who set them up. As implied in Figure 1.6, 'science' and 'governance' are just such institutions, creating a base for long-lasting organisations. In civil society there are a variety of formal and informal institutions and organisations, and it is wise always to find out what is 'there on the ground' instead of following an official 'map'. For example, in preparing for the SPICOSA SAF application to Loch Fyne in western Scotland, it was found that groups who might have an interest in the state of the loch included collectives of professional fishermen and shellfish farmers, loose associations or societies of recreational fishermen and birdwatchers, and the managers of the private company operating harbour facilities in the region (Tett *et al.*, 2011). It is important to catch this complexity, and so the utility of *institutional mapping* lies in the insights it provides into the institutions/organisations, and their economic, legal and social, interrelationships, relevant to particular coastal zone problems (Mette *et al.*, 2011). In chapter 4 (Figure 4.6) you can find one example of what an **institutional map** for clam fishing in Venice might look like.

Stakeholders, as has already been stated, are people who have an *interest* or *stake*, meaning a legal or moral claim or a right to participate in or be immune from, a particular aspect of the coastal zone, including the services it provides or the harm it may cause. Notice that such interests are always particular – they relate, for example, to the right to catch wild salmon or the right to prevent other people catching salmon in a certain reach of a river. Or they can relate to a firm's 'right to pollute' water in the form of a discharge permit, as well as the right of the public to a clean water environment. And stakes will always vary in terms of their strength or intensity or legal importance. As we saw in earlier sections of this chapter, some coastal zone problems can involve more socio-economic than ecological complexity, and **stakeholder-issue mapping** (Vanderlinden, 2011) therefore provides a tool for identifying the stakes and the categories of stakeholders who are legitimately concerned, and who therefore should be consulted, in relation to a particular coastal problem.

As already discussed, there are several ways in which scientific knowledge can be brought to the aid of coastal zone societies. In a 'top-down' approach, scientists enter in a dialogue with policy-makers at an appropriate level of governance, leading to improvements in the rules (i.e., directives, laws or regulations) for ensuring sustainability, or in their implementation. In the approach used during the SPICOSA project, an implementation of the Systems Approach Framework involved opening a **communications space** (Figure 1.6) amongst scientists, policy-makers, and stakeholders, so that the latter could contribute their knowledge of the social-ecosystem and voice their own interests. This is the process of deliberation, which is considered more fully in chapter 5. It is supposed to lead, initially, to the identification of one or more Issues, which are problems that have been agreed as being relevant to stakeholders, and so appropriate for further analysis. An Issue is a package, a task specification, in some ways like the brief agreed with an architect before a new house is designed and built. As a minimum it includes a statement of the ecological and socio-economic cause-and-effect chain from a human activity to an impact on human use of ecosystem services, an institutional map showing the formal and informal rules (including laws) that can facilitate or constrain solutions, an account of relevant stakeholder interests, and other elements in the analysis that will be addressed in the next chapter.

Conclusion

No matter how challenging the task of utilising science for the improvement of coastal policies may be, we must never forget the overall objective of all these efforts. At the foundation of this lies a strong conviction that by increased effort to represent and analyse coastal problems in a more coherent and systematic way, we can better secure the coastal ecosystems that are still intact, and restore or enhance those coastal systems that are no longer intact. Only through this can we secure a long-term sustainability of coastal ecosystem goods and services as a common objective.

Endnotes

[Introduction] Ideas about the flow of information from ecosystem to social system derive from the work of Niklas Luhmann, further considered in chapter 5.

[Travels with a salmon] This section is not to be confused with the famous essay by Umberto Eco, *How to travel with a salmon*. The challenges of travelling with a smoked salmon are of a different nature than the problems facing a wild salmon. The river Clyde in Scotland was once well stocked with salmon (Hughes & Nickell, 2009). Discharges of domestic sewage and industrial waste into the river during the nineteenth century reduced it literally to black and stinking. Its estuary, near Glasgow, became impassable by fish such as salmon. The UK Royal Commission on Environmental Pollution, taking note of cases like this in urban estuaries around Britain, concluded that one of 'two simple biological criteria for the management of estuarial waters' was the 'ability to allow the passage of migratory fish at all states of the tide.' After more than a century, and thanks to improvements in sewage treatment and discharge, the Clyde estuary is once again able to be classed as a salmonid water. For further reading about aquacultural controversies, *see* Naylor & Burke (2005) and Young & Matthews (2010).

CHAPTER 3

The Systems Approach

Paul Tett, Anne Mette, Audun Sandberg, and Denis Bailly

Systems

Billiards, snooker and pool are games played with coloured balls on a cloth-covered table. The balls constitute a **system**, which has a set of components (the balls), a state (the current position of the balls), and a set of rules for the physical interaction between balls (Newton's Laws of Motion). In addition, the game in which they feature has some rules about, for example, how the players take turns in striking the balls with their cue sticks. These rules can be varied (compare the game of English billiards with that of snooker), unlike the Laws of Motion, which appear to be fundamental properties of matter. Then there is the matter of players, who set the balls in motion. Do they belong to the physical system or to the compound system that includes the rules of the game?

The Solar System comprises a set of planets and other bodies orbiting about the sun and in some cases each other. Their motions can (mostly) be described by Kepler's Laws and further explained by Newton's Law of Gravitation. One of these planets is our own blue and white Earth, itself a set of systems. Amongst these is the biogeochemical exchange of carbon between organic matter and atmospheric gases. Occupying only a tiny fraction of the planetary surface is the Lagoon of Venice (Figure 3.1), but this too can be considered as – indeed, may actually be – a system, existing with boundaries defined by barrier islands, such as the Lido, and by the Italian mainland. The components of this social-ecological system include water masses, the organisms that live in the water or on the lagoon bed, the fishers who harvest some of these creatures, the built environment of Venice, and the interlocking institutions of city life.

We have now touched on most of the main properties of a system as defined by the **General Systems Theory** (**GST**) of Ludwig von Bertalanffy and listed in Table 3.1. Two of these properties need further explanation. *Boundaries* are what separate

3.1 The Lagoon of Venice, a false-colour picture taken by a satellite in 2001. The city of Venice lies on the central island.

Table 3.1 Properties of Systems.

A system:
– consists of parts and relationships or interactions amongst these parts;
– often contains *feedback loops*, which create *emergent properties* additional to those of the individual parts and relationships;
– has *boundaries* in space and time, which define system *extent* and *scale*; boundaries are permeable, in the case of an *open* system, impenetrable in the case of a *closed* system;
– has an internal *state*, which responds to internal dynamics and trans-boundary processes;
– can contain a *hierarchy* of sub-systems; emergent properties of one level appear as relationships at the next higher level; sub-systems can also, or alternatively, form a spatial pattern, leading to system *granularity*.

the system under study from the 'rest of the world' (Figure 3.2). In the case of *open* systems, stuff can pass across these boundaries. But the effects of the passage are asymmetric: what enters the system from outside, affects the system state. What leaves the system is supposed to have no effect on this outside. *Hierarchy* means that the main system includes sub-systems, as the Venetian lagoonal system is included within the large systems of Earth, and the Earth in the solar system.

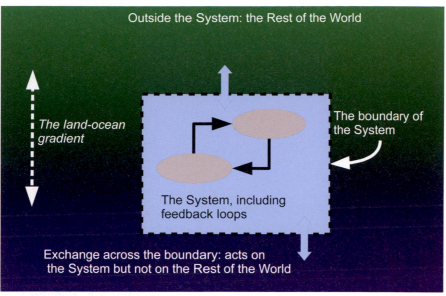

3.2 A coastal zone system showing boundaries and boundary conditions.

What we mean by the phrase, 'can be considered as – indeed, may actually be – a system' will require revisiting matters explored in chapter 2, concerning the difference between a **hard view of science**, in which a system is something that either exists, or doesn't, and a **soft view of science**, in which viewing some parts of the human or natural worlds as systems is a way of understanding these parts. In fact, the relationship between *hard* and *soft* views of systems lies at the heart of a Systems Approach, and it is that relationship that is central to the argument set out in this chapter. But before proceeding further into details of systems theory, we need to take a sideways step into philosophy.

More than *cogito ergo sum*

Remember that hot Mediterranean beach? Imagine yourself there, relaxing in a deck chair, eyes closed, scarcely disturbed by the sounds of children and the cries of gulls. How do you know that all about you is real? How do you know, as Descartes asked in the seventeenth Century, that you are more than a virtual object in a computer simulation? Of course, he didn't ask that, he asked his readers to wonder if all that they perceived was in fact manufactured for them by a meticulous as well as malevolent demon. 'How do I know that anything is real?' he asked. 'Well, at least I know that there is something doing the thinking, i.e., me. I cannot doubt that, whereas I can doubt the content of what I think.' Writing in Latin, he said: *'Cogito ergo sum'* – 'I think, therefore I am'. The distinction between what is perceived and what does the perceiving has occupied many of the working hours of philosophers, and more recently of neurophysiologists, who ask how electrical signals in nerves give rise to the stuff of thought, feeling and sensation. Without taking up any particular position on these matters, it is convenient to distinguish two worlds of phenomena:

❑ the real, physical, world in which things happen according to immutable laws such as those of Newton;

❑ the world of mental experience, which includes our thoughts and feelings and the need for *well-being* that is seen as the driver in economic conceptualisations.

Thus far, nothing philosophically unusual. However, there are phenomena that do not seem to fit easily into either of these two worlds. Many academics are familiar with being told to 'get real' instead of indulging in 'ivory tower speculation'. However, what is meant by 'reality', in such cases, tends to involve the earning of money. To which world does money belong? Not the physical world, because money is not governed by immutable laws: its value can change with inflation, or governments can decide to replace deutschmarks, francs or lire by euros. But not the mental world, either: I might imagine that I have a million euros, but that doesn't change the amount in my bank account. The twentieth-century philosopher Karl Popper proposed that we think of three worlds, the first two as already described, plus:

❑ …world 3, the world of information generated by the minds of sentient creatures, such as humans, and potentially comprehensible to other such minds; including communications between these minds, maps, models, songs and stories; and, more subtly, the institutions of society and the ideas, such as property or justice, associated with these.

World 3 and world 1 together form the reality experienced by the human minds of world 2. The difference between worlds 1 and 3 is that the laws of world 3 are not immutable. Human customs can change. To give a trivial example, billiards was first played outdoors, with mallets, and is now played indoors, on a table, with sticks. To give a more serious example, what is commonly thought of as something able to be owned, has changed with time: it once included persons. The hypotheses, theories, and models that summarise scientific knowledge of all three worlds are part of world 3. In relation to world 1, such understanding is either right or wrong, depending on agreed epistemological rules. The physical universe did not change when the theory of 'phlogiston' – a substance given off during combustion – was replaced by the theory of oxidation of combustible materials. But in the social world, a revolutionary theory can indeed lead to a revolution: people and society can be changed by the very act of understanding.

Hard and soft systems

General Systems Theory (*GST*; von Bertalanffy, 1968) approaches systems as entities that really exist. This does not mean that GST views systems as in principally 'organic', although in practice many actual systems are. It simply says that there exist sets of objects and processes that exhibit the features shown in Table 3.1. The idea of boundaries is particularly important: GST does not require systems to have *natural* boundaries, although some systems – such as the physical Lagoon of Venice – are so constrained. It is convenient for human observers, focusing on a small part of the total system that is the Universe, to define these boundaries. It means that a model of the Lagoon of Venice is not required to simulate the 'Big Bang' at the commencement of time, nor to link to events in distant galaxies.

In contrast, in **Soft Systems Methodology** (**SSM**; Checkland & Scholes, 1990), a system is a mental construct used in understanding the world. The original focus of SSM was in social systems, but (as we explored in chapter 2), the methodology can also be applied to hypothesis-building about the physical world. The crucial difference between a system under GST and a system under SSM is that the former is deemed to be real, even if our knowledge of it is imperfect, whereas the latter is merely one of many ways of conceptualising physical and social reality. In the case of the Lagoon of Venice, a visitor might be most interested in how to get about, and hence in the transport system, whereas the shell-fishers, who will be met again in later chapters, may be more concerned with the lagoon's natural ecosystem, which determines its productivity, and the city's system of regulations for exploiting the lagoon. Under SSM, systems have a purpose and are to be judged according to their utility in serving that purpose, whereas under GST, system descriptions – or *models*, as we'll come to call them – are to be judged in terms of their correctness.

SPICOSA's System Approach Framework combined features of both GST and SSM. It assumed that coastal zone social-ecological systems do really exist. They are known to be complex, but there is no need to describe much of this complexity, because the task is to focus in on a smaller set of relationships between particular human activities and their impacts. It is formal descriptions of such well-specified, cause-and-effect chains that can be used to explore management or policy options. Thus what is to be modelled is more like a *system* as conceived by SSM, and which we will call a **virtual system** or a *conceptual model*.

What we have to say in the next section applies equally to real systems and to virtual systems. However, insofar as we use language appropriate to mathematics to characterise system properties, it will be best to say that the focus is on virtual systems and their quantification by means of sets of equations. Nevertheless, for such modelling to be useful, there has to be a correspondence between the behaviour of mathematical functions and the behaviour of real systems. And one of the great discoveries of three millennia of science, is, that there is.

Feedback loops

Given our insistence on the importance of boundaries, it is sometimes convenient to see a system as a *black box* with an input pipe and an output pipe. What this means is that we take the contents of the system as given, and focus on the relationship between input and output. Figure 3.3 shows three simple possibilities, in the form of graphs. The input is quantified along the horizontal scale, the *x-axis*, and is called the **independent variable** or the **forcing variable**. The output is quantified on the vertical scale, the *y-axis*, and is called the **dependent variable**, because it is to a greater or lesser extent dependent on the input variable. In short, '*y* is a function of *x*'.

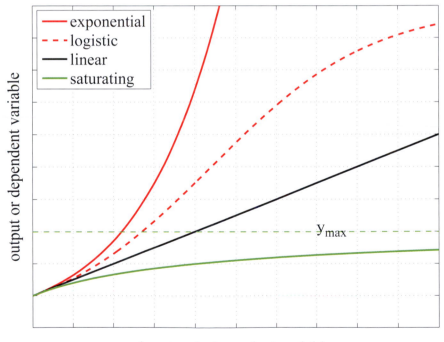

3.3 Simple input–output response curves.

In the simplest case, y changes in proportion to the increase in x. This is a **linear relationship**. An example is the small increase in sea level due to expansion of the oceans as they get warmer, or the small decrease in the number of **widgets** purchased as the price increases slightly. Such linear input–output relationships are, actually, rather uncommon in systems.

In the second case, y initially increases in proportion to x, but then levels off. This relationship might be generated in several ways. One way might be exemplified by railway journeys and national prosperity. As people's income (x) increases, they might want to travel more, but the total number carried (y) will depend on the amount of track and the number of engines and carriages. In such cases, it is common to refer to a *saturation curve*. The **parameters** of this curve, such as the greatest number of people that can be carried (y_{max}), could be changed by investment in railways, providing more trains and track. A second mechanism involves **feedback**. The system might, for example, consist of a hot water tank with a heater and thermostat. Initially the water temperature (the y-variable) is low. The heater is turned on (at time $x = 0$), and the water warms. But as the temperature approaches what has been pre-set, the thermostat intervenes to switch off the heater. Although this system achieves a desired end-point (e.g., water for a bath that is hot but not too hot), the mechanism is referred to as a *negative feedback* loop, for reasons we'll explore shortly.

The third case is that of a curve trending ever-upwards. One mechanism for this is the *positive feedback* that is involved in the exponential growth of populations of micro-organisms. More cells make even more cells, and so on, until space or nutrient or energy becomes limiting, in which case the curve may stop increasing and reach a plateau. This case is referred to as a *logistic curve*. Exponential and similar growth patterns are very common in systems; monetary inflation provides another example.

Feedback loops are important parts of many systems, and give them emergent properties. Although the desired water temperature is set by adjusting the thermostat, it is not a property of the thermostat but of the whole system of heater, thermostat, etc. There are several types of thermostat, one based on a breakable electrical contact operated by a metal strip that moves with temperature. This thermostat is set by adjusting the inter-contact distance at room temperature. Clearly, this thermostat knows nothing of its human owners' intentions: the final temperature emerges from the operation of all the system components, including something that we've not yet made explicit. This is the leakage of heat from the hot water tank to its surroundings, a flow through the system boundary. It is because the heater works against this loss of heat that the system is able to arrive at a stable temperature.

The operation of such a feedback loop can be formalised (Figure 3.4). This is a generic diagram, applicable to all examples of such loops. It is not a plan showing how to wire up a thermostat, but it reproduces in logic the way the heater–thermostat system works. In the causal structure shown in the Figure, the system variable (H) is disturbed by some tension (T) and then sends a signal to a sub-system or component (S) that reacts by stepping up its activity so as to bring the level of the H back within acceptable limits. This logic is called a negative feedback loop because the indicator signal must generate a response that is opposite in sign to that of the deviation of H from the 'desired' value.

We write 'desired' because the target value is indeed that wanted by the human owners in the case of a thermostat–heater system. But negative feedback loops are found in many systems that are not designed by humans: warm-blooded animals have mechanisms to maintain their body temperatures at a stable value above ambient, and many ecosystems and social systems seem to display this property of homeostasis (self-regulation).

In an evolutionary perspective, those system structures that are selected are those that best maintain the value of the H to compensate for variations induced by the T. Since the publication of the *Origin of Species* in 1859, it has become well known that species of organisms evolve as a result of selective pressures that eliminate unfit

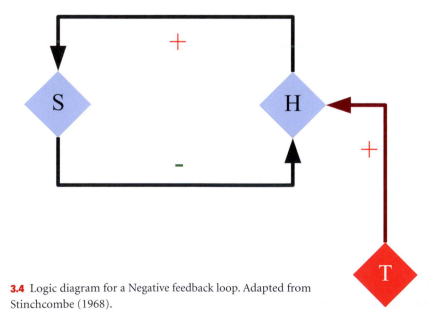

3.4 Logic diagram for a Negative feedback loop. Adapted from Stinchcombe (1968).

members, such as those that fail to control their body processes when challenged by external conditions. It is of course a requirement of Darwinian evolution by natural selection that *fitness* is controlled by heredity, so that better-adapted organisms can pass on their genes for fitness to their descendents. It is not so clear that there are analogous mechanisms in ecosystems or social systems, but it is an observed property of these systems that some outlive the conditions that generated them. This may simply be because systems that include negative feedback loops tend to survive, whereas those that include positive feedback loops don't. But there are additional possibilities, to which we will return in the next section.

As we have already mentioned, the opposite of homeostasis is a logical structure with positive feedback; i.e. it is self-reinforcing to the extent that a system might either explode or implode. In coastal ecosystems this is not an uncommon phenomenon, exemplified by blooms of blue-green bacteria in the Baltic Sea, or sea-urchins taking over a lush seaweed landscape. This is often the result when organisms have the ability to collectively modify the environment to benefit themselves, or when for some reasons their normal predators – for example, the sea-otter consumers of urchins – are absent. Indeed, it is often the relationship between predator and prey that provides negative feedback control: an increase in prey leads to an increase in predators that bring prey growth under control. But whatever the details of the cause, boom often leads to bust. Resources become exhausted – all the seaweed is eaten and the urchins starve. Or the abundance of blue-green bacteria provides a rich food source for a marine virus, which has its own boom, killing the bacteria as it does so.

Such booms and busts occur also in the human world. Periods of economic growth alternate with market crashes, and both are in part self-sustaining. Growth creates confidence; people spend; entrepreneurs increase production; and so on. Once confidence is lost, people stop spending, factories are closed, people put out of work; and so on. Both are examples of positive feedback loops, which some financial regulators try to oppose with negative feedback, for example increasing state spending during periods of economic recession.

Adaptation, Resilience, State Change and Panarchy

Negative feedback loops allow systems to survive changes in their external environment – i.e. in what lies beyond the system boundary. Another mechanism for survival is redundancy in components and for 'functional alternatives' when initial structures fail. Unconstrained coastal communities can use a variety of ecosystem goods and services, and so, for example, might switch from herring to cod if herring migration patterns changed. Many ecologists argue that biodiversity is functionally important in ecosystems, not because every species has a vital role at all times, but because there are replacements available when what are temporarily key species are lost because of disease or human actions. Genetic diversity within species provides the raw material for natural selection, allowing a species, as a system-like entity, to adapt to changed environmental conditions. What we mean to suggest here is that systems that include functional alternatives in both components and interconnections, and which include more negative feedback loops, are more resilient in the face of change in their environment.

Resilience is, here, viewed as the emergent system property that makes the system response less than proportional to a perturbation, and which allows the system to recover rapidly from such perturbation. In Figure 2.2 it is the property that keeps system state high and to the left of the 'cliff', despite increase pressure.

Still, there are situations where a self-reinforcing process continues to work until it reaches a stage that is irreversible – where a system collapses, or 'falls over the edge of the cliff'. Let's examine what that means in a little more detail. If Figure 2.2 is interpreted as the response function of a system 'black box', it says that as pressure increases (along the x-axis), system state (measured on the y-axis) collapses. To make a more concrete version of this diagram, we need an indicator or measurement variable allowing us to quantify **system state**. We'll return to this in a later section. Meanwhile, we need to be a little more sophisticated in understanding what has happened, and to return to the generic input–output functions in Figure 3.5. These are more complicated than those in Figure 3.3.

One of these response functions shows *chaotic* behaviour: the system jumps, often unpredictably, from one state to another (measured on the y-axis) as a function of either pressure or time (drawn on the x-axis). The final function also shows a complicated change in system state, but this occurs more deterministically as pressure increases. It seems to be a general property of sufficiently complicated, potentially homeostatic, systems, that they can have several (quasi)-stable states. Going over the cliff may simply be the transition from one state to another; the problem of return may simply be that of getting out of the new stable state. For instance, sustained overfishing of the cod stocks on the Grand Banks, east of Canada, has depleted the stock to such an extent that crabs and other shellfish take over the habitat and exclude the cod, as mentioned in chapter 2. We might expect such changes to have knock-on effects on the social system, with shell-fishers replacing cod-fishers.

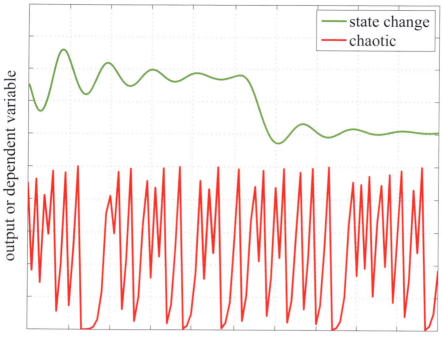

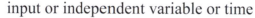

input or independent variable or time

3.5 Complex input–output response curves.

Scheffer and colleagues (2009) have suggested that systems that are about to change state exhibit the chaotic behaviour shown in Figure 3.5, with a big and erratic response to small changes in pressure. Such points of change are often called 'tipping points' (Gladwell, 2002), using the metaphor of a mechanical balance weighted down at one end. Sand is added to the other end, one grain at a time. All grains are equal, but one of them will be the one that causes the balance to tip. In a slightly different metaphor, a pile of small stones can be stable until one final pebble is added – and then the whole pile collapses in a little avalanche. But it is hard to predict exactly which pebble will start the collapse.

Words are important here. 'Change of state' is neutral. 'Collapse' implies something to be avoided. Implosions or explosions or tipping points can be seen as natural occurrences that reoccur at certain intervals. Thus they can have a near-cyclical character, as the creative destruction of one system releases matter, nutrients, energy, capital, talent – or whatever it takes to reorganise and start a new system – which might resemble the old system, but which might also contain qualitatively different elements or connections.

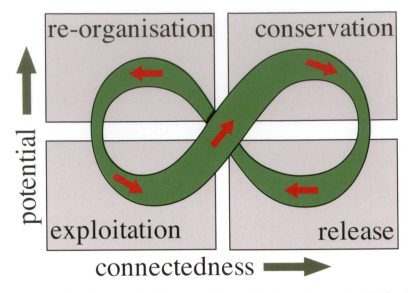

3.6 Panarchy: the adaptive cyclical character of an exploitation–release system. The horizontal axis represents complexity and connectivity, the vertical axis, labelled 'potential', can also be seen as representing natural capital in the case of ecosystems, social capital in the case of societies. The *conservation* (K) phase is that of maximum capital and complexity; the capital declines in the *release* (Ω) phase; the re-organization (α) phase is an alternative state with capital but little complexity. The *exploitation* (r) phase seems like another collapse, without capital or complexity, but it is from this that the next K phase grows by developing capital and complexity. Adapted from Gunderson and Holling, (2002).

The most famous presentations of this logic are found in the Panarchy discourse that attempts to connect ecosystem functioning with economic systems and the institutions that humans erect to govern the relation between these two sorts of system (Gunderson & Holling, 2002). In its simplest form this way of thinking connects the ecological maintenance and the economic exploitation of ecosystems in a non-linear way, involving four stages of a dynamic cycle (Figure 3.6). The stages are:

- ❑ The *exploitation* stage is one of rapid expansion, as when a population finds a fertile habitat in which to grow;
- ❑ In the *conservation* stage, the key feature is slow accumulation and storage of energy and material, as when a population saturates the **carrying capacity** of its environment, and stabilises for a time; ecologists expect that the fast-growing, *r-selected* species of the bloom phase give way to slow-growing *K-selected* species of a *successional climax*;
- ❑ The *release stage* occurs rapidly, as when a population declines due to a competitor, a calamity (like a forest fire), resource exhaustion or destruction, or dramatically changed conditions;

❏ The *reorganisation* stage can also occur rapidly, as when certain members of the population are selected for their ability to survive despite the competitor or changed conditions that triggered the release.

Some economic and social systems are thought to have this fundamentally cyclical character, e.g. entrepreneurial cycles and financial crises (Schumpeter, 1934). They would then share with the ecosystems the same stages of an adaptive cycle, which in more familiar language would be analogous to birth, growth and maturation, death and renewal. The dynamic character of all these systems would be determined by three different factors; two of which are represented by the axis of Fig 3.6. The 'potential' gives the limits to what is possible – i.e. the number and kinds of future options that are available. The 'connectedness' determines the degree to which a system can self-regulate, as opposed to being susceptible to influence by external variables. For instance, temperature regulation in warm-blooded animals has high connectedness because it involves at last five different physiological mechanisms. Ecologists argue that complex food webs and high bio-diversity allow ecosystems to respond with flexibility to changes in external conditions, and sociologists contend that multi-scale governance structures, with a strong civil society and with local as well as central decision-making institutions, act as buffers against the risk of social collapse.

System state

Now it is time to open the black box and consider how to specify the internal state of a system. Let's return to the game of billiards. Seen as a *hard* system, the state is determined by the positions and velocities of the balls. From a *soft* perspective, the current state might be that 'player B is winning', which is something that emerges from the physical position of the balls and the social rules of the game.

GST grew out of the scientific discipline of thermodynamics, for which the state of a gas provides a paradigmatic example. A given mass of a gas, such as helium, is made up of a certain number of molecules. Each molecule, like each billiard ball, has a current position and velocity. But any appreciable mass of helium contains billions of molecules, and defining the system state in terms of all these would be excessive. Let's suppose the system that we consider is that of the helium inside a toy balloon, so that the thin rubber membrane of the balloon provides the system boundaries. System state can now be defined by three emergent properties: the temperature, pressure and volume of the gas. Temperature is a result of the average speed of the gas molecules, which transfer energy as they collide with physical objects. Pressure results from the rate at which gas molecules bombard the walls of their container. Volume results from temperature and pressure: at a given temperature the balloon will expand until the gas pressure is exactly balanced by the tension in the balloon rubber membrane and the external atmospheric pressure. The 'Gas Laws' say that only two of these three properties are independent, for a given mass of gas. If temperature is kept constant, by allowing the balloon contents to remain in thermal contact with the surrounding air (so that the boundary condition of the system includes a flow of heat across the balloon membrane, as air molecules and helium molecules exchange momentum via the membrane), then knowledge of this temperature and the balloon volume allows the internal pressure to be calculated.

In this example, temperature, pressure and volume are the **state variables** that completely define the state of any system composed of an 'ideal gas'. Notice that all three are emergent properties of the underlying state of the system components (the molecules), and that we could replace any one of these variables by 'number of molecules' or 'mass of helium', because the Gas Laws prescribe the relationships amongst this larger set of variables. If such replacement is made, then one of the state variables becomes tightly linked to a decision about where to put the system boundaries (e.g. how big a balloon to use) or to an initial decision about how much gas to put in the balloon.

Now to more relevant systems: ecosystems and social-ecological systems. How do we find state variables for such systems? The *European Marine Strategy Framework Directive*, **MSFD**, uses the notion of 'qualitative descriptors for determining good environmental status'. There are 11 of these, and the European Commission's intention is that EU member states will identify a (small) set of variables that can be used to assess each qualitative descriptor. The fourth descriptor concerns food webs, and it has been suggested that amongst several useful indicators might be 'the productivity (production per unit biomass) of key species or trophic groups' and the abundance of 'large fish (by weight)' (European Commission, 2010).

In a somewhat similar fashion, the longer-established *Water Framework Directive*, *WFD*, uses the principle of ecological status, which is assessed by means of a number of quality elements. Amongst these is a 'biological quality element' for the 'composition, abundance and biomass of phytoplankton'. The United Kingdom, for example, interprets 'biomass' in terms of amounts of chlorophyll, the green pigment central to photosynthesis, and 'composition' in terms of identification and enumeration of planktonic micro-algae (Devlin *et al.*, 2007).

Because large fish are normally the subject of commercially important fisheries, and hence of great importance to humans, the 'large fish abundance' state variable might also be seen as a **natural capital**, analogous to financial capital in economics. Thus, the state of a social-ecological system could be defined by the values of all capitals, including natural stocks, **human and social capitals**, and **financial and durable capitals**. Alternatively, the system state could be defined, following Ostrom (2007) in terms of variables relating resource systems, resource units, governance systems and users or actors. Clearly, in moving from the WFD's aim to 'protect... and enhance the status of aquatic ecosystems' to Ostrom's 'make social-ecological systems...sustainable over time', we are moving from an ecosystem perspective to a more anthropocentric view of the social-ecological system. This perspective is explicitly maintained in the *Millennium Ecosystem Assessment* which attempted to place a money value on the services provided by ecosystems to humanity, and is the subject of a detailed example later in this chapter.

Management of systems

The thermodynamical approach to systems is the 'hardest' of 'hard' science: it provides a means of specifying system state without saying that one state is better than another. But humans want to appraise state in relation to some preferences. These might be a preference for a healthy ecosystem (because that delivers goods and services sustainably) and a healthy social system (because that is better for its members).

As we've discussed in chapter 2, the experiences of recent history seem to have taught twenty-first century humans that utopias are good only as daydreams. The perfectly 'good' society (however organised) and the undisturbed 'pristine' ecosystem (however recognised) are unachievable. As with J. R. R. Tolkien's Elvenhome in *The Silmarillion*, they are not harbours to which mortal men can steer. Instead, we as citizens can collectively wish to move towards a healthier state, and can require our navigators in governance to set a course that will take us in that direction – a guided voyage in state space, although without a final destination.

That is to say, we want to manage systems – which in the present case requires the management of human activities that have an impact on ecosystems and their goods and services. As we saw in chapter 2, this needs the development of policies that will take us in the direction of greater sustainability, equity and efficiency. Sustainability can be seen as preserving or increasing all capitals, including natural and social. Equity is dealt with in our method by stakeholder engagement. Efficiency can be quantified, and compared, in terms of the costs and benefits of the changes (or lack of change) resulting from the outcomes of several different scenarios (next section). The cost of moving from one system state to another needs to be proportional to the perceived improvement. In principle, both costs and benefits can be quantified in money terms, allowing them to be compared. Where the improvement requires human action, such as building a new sewage treatment plant, the money costs are easily known. But it might also be necessary to include **externalities**, hidden costs imposed on other people or in ecosystem destruction (for example, of the place where the treatment plant is built). Similarly, the monetary value of some benefits, such as enhanced fisheries, might be immediately obvious, whereas others might take some finding out. For example, Swedish people value the transparency of the water in the Himmer fjord, and in surveys express their willingness to pay a certain amount per visit in order to enjoy this intangible aspect of water quality (Franzén *et al.*, 2011). More generally, environmental economists have provided methods for evaluating all ecosystem services, as we will consider in relation to the Millennium Ecosystem Assessment.

We are getting ahead of ourselves. Let's look briefly at how the European Water Framework Directive deals with management. It requires that member states set up **programmes of measures** to achieve the Directive's aim of making a framework that 'prevents further deterioration and protects and enhances the status of aquatic ecosystems'. It requires that water (quality) status be monitored in terms of the quality elements set out in the Directive's Annex V, which includes the 'phytoplankton biological quality element' already mentioned. Management requires the use of monitoring both to identify what specific measures are needed as part of the programme of measures, and to track the effectiveness of the programme in moving aquatic ecosystem status towards the goal of maintained or improved quality.

The WFD is an example of policy-making at the *constitutional level* of governance: it provides a framework for rule-making (at the *collective level*) in each member state of the EU. At the *operational level* of devising and implementing local programmes of measures, the environment management agencies charged with this task have the challenge of identifying the best measures to remedy particular effects of human activity in particular rivers, estuaries or coastal waters.

That identification might be done on the basis of the experience of ecological experts, or by a try-it-and-see approach. Our argument in this book is that the most efficient and safe way to do this is to build models that embody the main characteristics of the relationship between a human activity and an impact on social-ecological system state, and to use these models to explore the likely outcomes of a number of management options.

Scenarios and options

If a water body is in danger of becoming eutrophic, and hence suffering damage to human use or enjoyment, the appropriate remedial measure is reduction in the inputs of nutrients – compounds of nitrogen and phosphorus. If we return to the view of systems as black boxes with inputs and outputs, nutrient loading is the input, and changes in ecosystem services due to eutrophication are the output. How to turn down the input tap? Engineers might suggest building things: a better sewage treatment plant, or a longer discharge pipe to take the sewage further out to sea, where it will be dispersed more effectively. In fact, there might be another option, which does not fit so neatly into this input–output formalisation: install a mussel farm (Figure 3.7). The shellfish eat phytoplankton that has grown using the nutrients, and so harvesting the mussels leads to removal of some nitrogen and phosphorus (Lindahl

3.7 Scenarios on a flipchart.

et al., 2005). This changes something inside the black box, so that the ratio of output (trophic state of the system) to input (nutrient loading) is altered.

At the policy level, a higher level of complexity arises. If, for example, a state's regulations are to change to allow shellfish aquaculture as an alternative to conventional sewage treatment plants, there are matters that bear on their choice. The industries that build and operate sewage treatment plants are distinct from those involved in aquaculture and the marketing of its products. Economic activity might be displaced between the two sectors: what might be the social consequences? Would the mussels be safe to eat? Would the public think that they were safe to eat? Suppose that there were to be changes in factors beyond the control of state government – increases in the price of energy needed to run a sewage treatment plant, or a developing global scarcity of protein? What then?

Thus it is often the case that the available options are complex, and are to be implemented over a number of years, in a future which is itself not defined. For example, the cheapest site to build a sewage treatment plant is usually downhill from the city that produces the sewage, because this allows sewage to flow under gravity and without the need for pumping. However, if the plant is built near the seashore, it might be at risk from increasing sea levels resulting from global warming. So there is a need to forecast sea level during the operational lifetime of the plant. But, it turns out, this is in itself a complicated matter, because it depends on collective choices made by humanity about emissions of greenhouse gases.

It was in order to explore the likely outcomes of such choices that the global community of scientists involved in making such forecasts began to think in terms of scenarios. According to the International Panel on Climate Change (IPCC), a *scenario* is 'a coherent, internally consistent and plausible description of a possible future state of the world'. The Panel has designed four sets of scenarios, based on four *storylines* about possible futures, each with different implications for climate change (Table 3.2).

Such global scenarios are to be seen as imposed on a local coastal zone system, or the 'virtual system' distilled from it, from outside. They are part of 'the rest of the world' in Figure 3.2, and, in chapter 4, it will help to think of them as **boundary conditions** for the equations of the model. They are outside the control of the social-ecological system under consideration, whether they be global warming or the stipulations of the European Urban Waste Water Treatment Directive.

In contrast, there are management options that are under local control, such as the choices of building a larger sewage treatment plant or installing a mussel farm.

Table 3.2 IPCC 'storylines and scenario families', from IPCC-TGICA (2007). These scenarios are used to provide time-series of inputs of greenhouse gases to models that predict future changes in climate and sea level.

A1: a future world of very rapid economic growth, global population that peaks in mid-century and declines thereafter, and rapid introduction of new and more efficient technologies.
A2: a very heterogeneous world with continuously increasing global population and regionally oriented economic growth that is more fragmented and slower than in other storylines.
B1: a convergent world with the same global population as in the A1 storyline but with rapid changes in economic structures toward a service and information economy, with reductions in materials intensity, and the introduction of clean and resource- efficient technologies.
B2: a world in which the emphasis is on local solutions to economic, social, and environmental sustainability, with continuously increasing population (lower than A2) and intermediate economic development.

The first of these might be modelled by a simple reduction in the input to the system black box, whereas, as we wrote above, the second might require a design change inside the virtual system.

Let's move on to something trickier. What happens to the ecosystem if the mussel production is increased? What happens in the local market for shellfish if changes in mussel production and mussel consumption are out of balance? The capacity to simulate the first should be straightforward to include in the ecosystem model. Simulation of the second might be done with a simple economic model, using micro-economic principles and two parameters, quantifying willingness to pay for mussels, and willingness to eat mussels. A determined modeller might construct a deterministic model that would predict everything, including the values of the two 'willingness' parameters, for the scenario of installing a mussel farm for sewage treatment. However, it might be easier to treat the values of these parameters as part of subsidiary scenarios within the main 'install a mussel farm' storyline. Suppose that a survey suggested that there would be a higher willingness to pay if people believed that mussels improve your love life, then this could be fed into a greater value of the 'willingness to pay' and 'willingness to eat' parameters in one sub-scenario. The other would assume no change from the default values in these willingnesses.

Such use of scenarios can be thought of as 'what-if-ery': 'what happens if we do A, or B, or C'. Here A, B and C are the inputs to the black box of the virtual system: the management or policy choices, and the externally imposed components of scenarios. The 'what happens' are the outputs. In fact, scenario can be used to label either the inputs or the resulting outputs. For example, in reporting model results to stakeholders, a presenter could refer to 'the possible future of what will happen to the local economy and the water body, if people would start eating mussels like crazy'.

The Millennium Ecosystem Assessment

The Millennium Ecosystem Assessment (**MA**) was called for by the United Nations Secretary-General Kofi Annan in 2000. Initiated in 2001, the objective of the MA was to assess the consequences of ecosystem change for human well-being and the scientific basis for action needed to enhance the conservation and sustainable use of those systems and their contribution to human well-being. The MA has involved the work of more than 1,360 experts worldwide. Their findings, contained in five technical volumes and six synthesis reports, provide a state-of-the-art scientific appraisal of the condition and trends in the world's ecosystems and the services they provide ... and the options to restore, conserve or enhance the sustainable use of ecosystems.

We will use the MA to illustrate two sets of tools. The first concerns the valuation of the services that ecosystems provide to humans. This, we have argued, is one way to define the state of social-ecosystems. The second set of tools are those concerned with the formulation and use of scenarios.

The MA developed four scenarios with which to explore the potential dangers to human well-being during the next hundred years. These scenarios (Table 3.3)

Table 3.3 Scenarios of the Millennium Ecosystem Assessment. From MEA (2005). In all the scenarios, human pressure on ecosystems was supposed to increase for at least the first fifty years. The forces of change taken into account were: habitat change (changes in land use, physical alteration of rivers or extraction of water from rivers); over-exploitation; invasive species; pollution; climate change.

Global Orchestration: 'a globally connected society that focuses in global trade and economic liberalisation and takes a reactive approach to ecosystem problems ... Also takes strong steps to reduce poverty and inequality and to invest in public goods such as infrastructure and education.' Highest economic growth and lowest population in 2050.

Order from Strength: 'a regionalised and fragmented world, concerned with security and protection, emphasising primarily regional markets, paying little attention to public goods, and taking a reactive approach to ecosystem problems.' Lowest economic growth and highest population growth.

Adapting Mosaic: 'regional watershed-scale ecosystems are the focus of political and economic activity. Local institutions are strengthened and local ecosystem management strategies are common; societies develop a strongly proactive approach to the management of ecosystems.' Economic growth starts low but increases; high population in 2050.

TechnoGarden: 'a globally connected world relying strongly on environmentally sound technology, using highly managed, often engineered, ecosystems to deliver ecosystem services, and taking a proactive approach to the management of ecosystems in an effort to avoid problems.' High economic growth; population in 2050 in the middle of the scenarios.

were constructed using the pooled opinions of experts on the possible futures of ecosystems, ecological services and human well-being, together with the results of global models for the main pressures on ecosystem services. Two different assumptions were used in relation to each of the topics of globalisation and ecosystem management. With respect to globalisation, dynamics would be either regional or global. With respect to environment, management would be either proactive or reactive.

The valuation tools included: the identification of ecosystem services to humans (Table 3.4); the money valuation of the natural capitals required by those services; and the money valuation of the services themselves. To those who object that, in the words of Oscar Wilde, they know 'the price of everything and the value of nothing', the authors of the Assessment argue that the:

> conceptual framework for the MA places human well-being as the central
> focus for assessment, while recognizing that biodiversity and ecosystems
> also have intrinsic value and that people take decisions concerning ecosys-
> tems based on considerations of well-being as well as intrinsic value. ... The
> MA plans to use valuation primarily ... as a tool that enhances the ability
> of decision-makers to evaluate trade-offs between alternative ecosystem
> management regimes and courses of social actions that alter the use of
> ecosystems and the multiple services they provide.

Table 3.4 MEA categorisation of ecosystem services. From MEA (2005).

Category	Examples	Examples of how valued
Provisioning services ('goods')	Food, fuel, fibre, fresh water	Market prices
Regulating services	Regulation of air quality, climate, runoff quantity, erosion; control of pests and diseases; waste absorption capacity; pollination; amelioration of hazards	Replacement costs, e.g. building sea-wall to replace mangroves as defences against tsunamis
Cultural services (non-material benefits)	Cultural heritage sites, aesthetic values, and value for education or recreation	Willingness to pay; premium on prices – e.g. higher value of houses with sea view
Supporting services (needed for the production of other services)	Soil formation, primary production, nutrient cycling, water cycling	Costs resulting from damage to these services; restoration costs

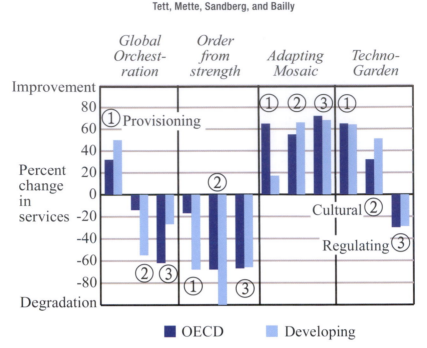

3.8 Results of the scenarios of the Millennium Ecosystem Assessment after Fig 5.3 in MEA (2005). The circled numbers 1, 2, 3 refer to categories of ecosystem services: see Table 3.4.

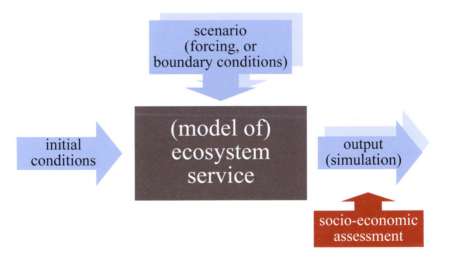

3.9 Augmented black box. In a simple *black box*, the system is considered as a box with hidden contents and input and output pipes. In this augmented black box, the virtual system is supposed to be that responsible for a particular ecosystem service. One of the inputs to this box is always the same set of initial conditions. The other input is a scenario, and the diagram suggests that several scenarios can be applied in sequence, each resulting in an output – or model simulation, to use the language of the next chapter – that gets appraised in socio-economic terms.

Figure 3.8 summarises the results of applying these valuations to estimates of changes under each of the scenarios, distinguishing between 'rich' (**OECD** member) and 'poor' (developing) countries. The results don't appear explicitly in terms of money, but as percentage of change (relative to the present day) in each service under each scenario. That is to say, the valuation is used as a navigational tool rather than as a price-tag for a service. The methodology used by the MA was complicated, but can be understood in terms of the 'black box' input–output machine of Figure 3.9.

The Systems Approach Framework

This chapter has viewed social-ecological systems in relation to General System Theory's definition of system, with some nods in the direction of 'systems as heuristic devices' of Soft System's Methodology. A Systems Approach is holistic and contrasts with the focus on parts that characterises an Analytical Approach. An example of the latter is the method used by economists to construct a supply- or demand-curve for a particular manufactured good or service, which assumes that this exists independently of the remainder of society or the state of the natural resources used (ultimately) to make the goods. Analysis has proved a powerful method for scientists during the last four centuries. However, it is less suitable for dealing with entities that display complex and interconnected behaviour.

The problem with an integrated approach is its implication that everything must be taken into account. What we are arguing in this book is that this difficulty can be avoided by the combination of GST and SSM: by the description of a virtual system that distils the main features of the behaviour of the real social-ecological system relevant to a particular chain from human activity by way of change in ecosystem state to impact on goods and services used by humans. Our Systems Approach Framework, or SAF, is scalable. On the largest scales, which include collective and constitutional levels of governance, the virtual system should include those components of the social-ecological system. On smaller, more local scales, the size of most of the examples that we present, the virtual system scales down to a model that in most cases comprises predominantly ecosystem properties, perhaps with some micro-economics. Larger-scale features become part of boundary conditions, and hence of the scenarios used to drive the models. Alternatively, they can be dealt with as parts of the social, economic and legal frameworks used to appraise the results of simulations made with the models.

In the next chapter we open the black box and dig into the details of model-making.

EndNotes

[More than cogito ergo sum] Popper (1978) explains his 'three worlds'.

[Feedback loops] Logical structures are world 3 entities, ways of reading patterns into real world (1 and 3) behaviour. As with all explanations, there is always a danger of trusting too much in the logics, and in using them to claim a natural tendency to 'balance' in pristine nature (Cuddington, 2001). Interactions amongst multiple types of organism can lead to greater stability in ecosystems (Ives and Carpenter, 2007). Removal of top predators has been shown to cause 'trophic cascades' (Pinnegar *et al.*, 2000; Jackson *et al.*, 2001). However, natural systems can exhibit strong natural variation, and detailed studies at species level often reveal greater complication than implied by simple feedback loops or even the cascade logics shown in the reviews mentioned above. The relationship between sea-otters, sea-urchins and seaweeds is one of these trophic cascades, described in the reviews and in many detailed studies, including North & Pearse (1970), Duggins (1980) and Sala *et al.* (1998). Concerning cyanobacterial ('blue-green algal') blooms, *see*: Stal *et al.* (2003); and Suttle (2000) concerning viral infections. Stiglitz (2010) describes the recent collapse in financial confidence and its consequences.

[Panarchy etc] There has long been a debate about the role of biodiversity in ecosystem stability or resilience (McCann, 2002). In species evolution, selection operates on phenotypes, which result from genotype–environment interactions, but heritability means that selection of certain phenotypes will change the frequencies of some alleles in the species gene pool. Benincà *et al.* (2007) has demonstrated chaos in an experimental plankton community. One of the 'messages of ecology' in the highly readable text by Charles Krebs (1988) is that 'communities can exist in several stable configurations'. Concerning 'regime shifts' in aquatic ecosystems, *see*: Scheffer and Carpenter, (2003). Concerning changes resulting from cod overfishing, *see* Frank *et al.* (2005) and Link *et al.* (2009). A more complex version of the Panarchy diagram can be found at: http://www.resalliance.org/index.php/panarchy. It contains a third dimension, for scale, and underlines the importance of inter-scale connectedness for both conservation and recovery. Most of the arguments about connectedness relate to resilience, which determines how vulnerable a system is to unexpected disturbances and surprises. But resilience is usually understood as the ability to resist, or adapt to, change, without losing essential characteristics. The Panarchic phases of collapse go beyond resilience; they are the birth of a new system from the ashes of the old, a system that might have quite different properties. Thus these phases correspond to a state change in the more standard discourse about systems.

[System state] Strictly speaking, temperature is proportional to the average translational kinetic energy of the gas molecules, and gas pressure results from momentum transfer as molecules collide with the walls of their container. The Gas Laws comprise laws ascribed to Boyle, Charles, Gay-Lussac and Avogadro. They are supposed to apply exactly to 'ideal gases', for which Helium, at moderate temperatures and pressures, is a good approximation. The Helium molecule is unusual in that it consists of a single atom.

[Scenarios] Much of the literature about scenarios tends to refer to 'visions' of the future, but 'vision' can suggest a rather idealistic picture. These dreams of possible futures may also be nightmares, and the word 'scenario' can imply either. The IPCC document (IPCC-TGICA, 2007) from which the quote is taken, distinguishes *storyline:* 'a narrative description' and *scenario,* a projection 'of a potential future, based on a clear logic and a quantified storyline'. One storyline can generate several scenarios. They add that 'a set of scenarios is often adopted to reflect, as well as possible, the range of uncertainty in projections'. We don't know of evidence that mussels are indeed good for your love life, but it seems likely that, compared with a diet of steak, they have beneficial effects on blood cholesterol. *See* what the aquaculture industry has to say: http://www.shellfish.org.uk/shellfish_healthy_eating.htm.

[Millennium Ecosystem Assessment] Oscar Wilde gave the words about price and value to Lord Darlington in the play, 'Lady Windermere's Fan', 1882. The 'rich' and 'poor' countries are so judged in relation to the value of their GDP per head. *See* http://www.oecd.org/ and http://en.wikipedia.org/wiki/List_of_countries_by_GDP_(PPP)_per_capita. Gross Domestic Product is based on money economies and takes no account of other aspects of well-being, including non-market access to ecosystem services. But it is a conventional way to compare states. Valuation as a navigational tool: even this usage is open to criticism: for instance, a valuation of an ecosystem service using 'willingness to pay' will likely give more influence to the views of rich people, who are willing to pay more. The first quote is from the MA website, http://www.maweb.org/en/About.aspx, from which many relevant documents can be downloaded. A good summary of the MA is given in Millennium Ecosystem Assessment (2005). The second quote is from Millennium Ecosystem Assessment (2003).

[Systems Approach Framework] Concerning the integrated approach, *see* Moss (2008), who discusses the extent to which the WFD's intentions of managing the 'reality of reasonably distinctive, integrated systems...are being undermined for ostensibly political convenience through processes of redefinition and limitation of characteristics measured.'

CHAPTER 4

Modelling coastal systems

Paul Tett, Maurizio Ribeira d'Alcalà and Marta Estrada

Introduction

The Systems Approach Framework includes a step called System Design. What is to be designed is a model of the coastal zone under study – or, to be more precise, a model of the virtual system corresponding to the parts of the coastal zone that are relevant to the environmental problem that is to be addressed. By the word **model** we mean, a simplified representation of the essential or dominant features of relationships amongst components of real systems, used to (i) increase and promote understanding of the real system, and (ii) simulate the behaviour of the real system under particular scenarios.

This definition needs some unpacking. It will help to travel back two and a half centuries, to the Scottish Enlightenment and attempts to understand, not only how the world worked, but also how humans came to know how it worked. Here is a quote by a famous eighteenth-century author:

> Systems in many respects resemble machines. A machine is a little system, created to perform, as well as to connect together, in reality, those different movements and effects which the [maker] has occasion for. A system is an imaginary machine, invented to connect together in the fancy those different movements and effects which are already in reality performed.

The author was, in fact, Adam Smith, writing not about economics but about the history of astronomy, and discussing various explanations, or 'systems', of the motion of the planets across the sky. Consider the orrery in Figure 4.1, a clockwork model of the solar system. Clearly, the model is not the real thing, nor is it an exact miniature of the real thing. It is a mechanical device that simulates orbital properties sufficiently well to explain the paths taken by the wandering stars in the sky. Indeed, as Smith argued, it was not the clockwork that was 'the system', but what was in the heads of astronomers. Once this had been written down as mathematical equations,

4.1 Picture of a clockwork orrery. The word 'Orrery' comes from the title of the 4th Earl of Orrery (patron of the inventor, George Graham), and was first used in the early 18th Century for 'an apparatus showing the relative positions and motions of bodies in the solar system by balls moved by a clockwork' (Meriam-Webster online dictionary).

it became possible to replace clockwork models by sequences of calculations; and, in the second half of the twentieth century, to carry out these calculations using computer programs.

This chapter deals with two sorts of models: those inside human heads, or externalised as maps or diagrams, which we will call *conceptual models*; and those involving the solution of sets of mathematical equations, which we will call *simulation models*. We shall say something about how simulation models are made – as far as possible eschewing formal mathematics – and about how models can be used in coastal zone management, within a Systems Approach Framework.

Problems in the Lagoon of Venice

In the tourist brochures, Venice's Grand Canal throngs with gondolas, and the beaches of the Lido are packed with sunbathers. But the Lagoon of Venice seems a less promising place on a misty winter's morning. The water taxi from the airport moves slowly amongst muddy banks topped by bleached reeds, then picks up speed along routes defined by lines of posts in the sea. It is grey and cold. The occasional bird, disturbed by the boat's passage, flaps sullenly away from its post. From time to time, other small boats appear in the fog, working gear over their side. Then, at last, a dull grey smudge begins to resolve itself, as if a battleship rests on the mud, its upper works above the water; and then it is clear that the shapes are those of the domes and towers, the wharfs and buildings, of the city of Venice, rising directly out of the sea.

When it was first settled, the Lagoon of Venice was a patchwork of shifting banks of mud amidst saline channels, protected from the open waters of the Adriatic Sea by the larger sandbanks that form the Lido. In order to live here, people had literally to make land, stabilising the sand and mud banks with wooden rafts, held in place by piles driven through into the underlying clay. But it was worth it, for the protection from enemies and for the fish that could be caught in the lagoon's waters. In this place, from the ruins of the Western Roman Empire, emerged the Republic of Venice, whose ships dominated trade in the eastern Mediterranean for hundreds of years.

In Venice, there are no cars and indeed few ways wide enough for a motor vehicle. Instead, goods and people are ferried by boat. The sea is everywhere: the Grand Canal divides the island into two, and smaller canals penetrate every district. There is a price to be paid, of course, for the sea would like to reclaim the city, and so maintenance must be constant. It helps that the Mediterranean Sea is largely tideless, so that water level typically varies only just enough to flush the canals. But there is a long-term threat: Venice is sinking, as the weight of its buildings presses down into the mud and underground fresh water is sucked up to quench the thirst of its citizens and industries. Sea level is rising, due to thermal expansion of the oceans and the melting of polar ice. And there is a medium-term problem. Although the astronomical tide – that due to the pull of sun and moon – is small in the Mediterranean, whose basin is the wrong size to resonate to these pulls, meteorological conditions over the Adriatic Sea can raise or lower sea level by a metre or more at the Sea's head. The result is an *acqua alta*, a 'high water', causing partial flooding of the lower parts of Venice.

If you are thinking of visiting Venice and want to know whether there will be such a flood during your visit, there is a website that carries water-level predictions for the next few days. Behind this website is a computer program implementing a mathematical model – in fact, two models, one for the weather (to predict changes in atmospheric pressure over Italy and the Adriatic) and one to calculate the response of the sea to a particular distribution of pressure. Such models take today's observations of the state of the atmosphere and ocean and calculate how these states will change during the next few days, with the intention of making a single set of predictions. Although the models are complicated, the use of their results is simple.

In contrast, typical SAF models are simpler but have a more complex use. They evaluate scenarios: they can be used to explore 'what-ifs'. As chess players think through the consequences of possible moves before selecting a piece, SAF models evaluate the ecological, economic and social consequences of a potential management choice in order to allow stakeholders and managers or policy-makers to consider outcomes before committing funds or political capital to one particular management plan.

The subject of SPICOSA's study of the Lagoon of Venice was not an external threat, like the risk of a storm surge or the consequences of the global rise in sea level, but an internal conflict over the use of a provisioning service provided by the lagoon. The Manila clam, *Tapes philippinarum*, was purposely introduced here in 1983. By the 1990s it was sufficiently abundant to support a dredge fishery, which damaged the bed of the lagoon and led to clashes with long-established fisheries using fixed gear. That wasn't the only problem. Extensive re-suspension of sediment by the dredges might threaten the physical state of the lagoon. Pollution by sewage and toxic industrial waste contaminated shellfish and rendered some of them unsafe for human consumption. Nutrient enrichment of the lagoon's water by sewage and by agricultural drainage from the mainland caused problems of algal blooms, but on the other hand increased the food supply for filter-feeding shellfish.

The solution that has begun to emerge involves licensing restricted areas of the lagoon for the clam fishery. However, there are several possible strategies for exploitation, including a move to aquaculture, in which some of these areas are seeded with small clams obtained from nursery grounds. That led to the Issue studied at SPICOSA study site 15, which concerned the sustainable management of the clam fishery in the lagoon. A model was developed 'to represent the essential dynamics of major ecological, social and economic clam farming system components to project the consequences of implementing alternative management policies and to address the ecological and social carrying capacity.' (Melaku Canu *et al.*, 2011). We'll see that the cause-and-effect chain implied by the Issue, and simulated by the model, runs (in its simplest form) from phytoplankton through clams to the income and jobs that the clam fishery sustains in Venice.

Shellfish, phytoplankton and water exchange

The cultivation and harvesting of shellfish was an important topic at more than a quarter of SPICOSA's sites, and so we'll use the topic of shellfish aquaculture as the context for our examples of making and using models. As is known from archaeological excavations, coastal humans have long collected and eaten shellfish from the intertidal shore or shallow sea. Boats and dredges allow more effective harvesting of seabed animals, but, as the Venice lagoon case study shows, risk over-exploitation and habitat destruction. Sustainable aquaculture needs a reliable source of young shellfish, a place to grow and harvest them without excessive impact on the environment, and a sustainable source of food for the growing shellfish. The Venetian clam fishery relies on natural settlement of the pelagic larvae into nursery areas, and thus the maintenance of a wild stock that is protected from exploitation. 'Even if clam harvesting now is allowed only within dedicated lagoon areas, a rational integrated management of the resource has not been achieved. Social issues persist, including natural clam seed provisions and conflicts among fishermen to obtain more productive areas' (Melaku Canu *et al.*, 2011).

Bivalve molluscs, including mussels, oysters and clams, feed by exposing their gills to a flow of seawater, from which they filter food particles. In the case of clams, which lie partly buried in mud, the flow is sucked in through a muscular tube, a siphon. The best food is provided by the tiny floating algae that make up the phytoplankton, together with the pelagic protozoans associated with them. Both are rich in proteins, fats and phosphorus-containing compounds as well as energy-containing carbohydrates. But phytoplankton are seasonal, scarce in winter, when daylight is scanty, and, under pristine conditions, usually scarce in summer because of a shortage of the dissolved nitrates and phosphates that the algae need for growth. Coastal waters are often rich in particulate organic matter, **POM** for short. This is less rich in proteins, etc., but can sustain the energy needs of shellfish when other food is scarce.

In order to model the cultivation and marketing of filter-feeding shellfish, it is necessary to understand these and some additional effects, and to describe:

❑ the ability of individual shellfish to filter food from seawater and to use it for growth;

❑ the ability of pelagic micro-algae to grow in response to illumination and a supply of nutrients;

❑ the effect of water movements in bringing phytoplankton and POM from other areas inside the lagoon and from the external waters of the Adriatic Sea;

❑ the effect of populations of cultivated shellfish on phytoplankton and POM, which depends on the growth of individuals, natural mortality, and the seeding and harvesting schedules imposed by farmers;

❑ the effect of processes that occur outside the lagoon, but which influence the amount of clams that can be safely and conveniently sold in a sustainable way in the market; they include the potential pollution of the lagoon by poisonous substances or pathogenic bacteria, the harvesting cost (fuel, wages, etc.) and the costs for monitoring and prevention of pollution.

The last group of processes are tightly linked with the socio-economic part of the issue, and bear on the operation of clam fishing businesses as economic and social entities. Some of them may enter an investigation by way of scenarios (chapter 3) rather than as simulated processes, or through the socio-economic appraisal and interpretation (chapter 5) of the model's simulations.

Designing a system and formulating a conceptual model

That list of bullet points is the beginning of a system design. Bearing in mind that systems are hierarchical, it may be assumed, for the moment, that the details of how phytoplanktonic algae, or clams on the seabed, grow, are out of sight, inside conceptual boxes labelled 'phytoplankton' or 'clams'. Then the boxes can be connected up, and others added, as exemplified in Figure 4.2. This is a first conceptual model of the system that will become the subject of a simulation model, and much needs to be done to improve it before then. At this stage it is no more than a chain: it has none of the complexity and feedbacks that we expect to find in a social-ecological system.

The first step in improvement is to ask, what does this drawing mean? For example, there is an arrow running from 'phytoplankton' to 'clams'. What does that arrow signify? Just that phytoplankton have some effect on clams? Or, more precisely, that the more phytoplankton, the more clams? Actually, it should signify something more complex: that the micro-algae of the phytoplankton (as a result of their photosynthesis) provide a source of food to clams (allowing them to grow). That's a statement that suggests a need to think accurately about what we mean by food, and how we will quantify it, so that at a later stage we can formulate a mathematical relationship between food supply and clam growth.

As anyone knows, food in the market is generally sold by weight. Clams are sold wet, so the price we pay is per kilogram of wet clams. This includes the shell, which is not eaten. What we do eat is the flesh, of which we digest the different compounds (proteins, fats, etc.) for our nutrition. Those compounds are made of different elements (carbon, nitrogen, etc.) which occur in more or less constant ratios. Since

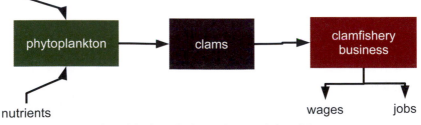

4.2 Simple conceptual model: phytoplankton, clams and clam fishery.

the growth of any organism is an increase in its biomass, and thus an increase in its content of carbon, nitrogen etc., one possibility is to choose the amount of nitrogen (N) as a working unit. This would allow the same unit to be used for the dissolved inorganic nutrients in the lagoon, for the phytoplankton, which collectively assimilate these nutrients to grow, and for the clams. The only information we would need is the ratio between the wet weight of a clam and its nitrogen content. The N content of an adult clam turns to be 7 mg per gram of wet weight (Nizzoli *et al.*, 2006). Thus we have a link between the units used by the 'economical' side of the business (the wet weight) and the nitrogen units used for analysing the 'natural' side of the business.

Next question: does the arrow show the nitrogen gained by a single clam when it eats a single alga? Or the nitrogen gain in 24 hours from all the algae in all the water that the clam has pumped over its gills in that time? Or should we be thinking that we are dealing with all the clams in a particular fisheries concession? Or all the cultivated clams in the whole lagoon? To keep things simple, we might conclude that each of the three boxes refers to the entire lagoon, and that all fluxes refer to a time period of one complete day.

But that simplifying assumption leads to other complexities. Neither clams nor clam fishery businesses are uniform: some are big, some small, some efficient, some less so. So whatever mathematical formulation is used, it must take account of this variability, perhaps by working with average properties, perhaps in more complex ways.

The SPICOSA team working on the Lagoon of Venice began to deal with this challenge by building a conceptual model of the physiological processes that control the life of a single clam (Figure 4.3). The main processes are the getting of food from material in seawater, which is sucked into the buried clam through its siphon, and the use of this food to make clam flesh. Some of the food material is used in respiration, to make the energy needed, for example, to power the siphoning and filtering of water. In addition to the micro-algae of the phytoplankton, clams also suck in and digest particulate organic matter, or POM, floating in the sea as the result of the death and decay of seaweeds, sea-grasses, etc., or even discharged in sewage. They also gain nutrition from other little single-celled organisms, such as a protozoa and bacteria, associated with this POM. Some of the clam's food, including the little silicified cell walls of some of the micro-algae, is indigestible, and is expelled as faeces.

Once formulated into equations, this core part of the description could be called a **Dynamic Energy Budget**, or **DEB**, model, because it balances food input (which can be considered as energy as well as nitrogen) against energy use in growth and respiration – and, not shown in the diagram, the use of energy (and nitrogen) to make eggs. But the clam model for the lagoon needed to take account of some more processes, and so the modellers also considered the reproduction by egg release, hatching and generation of larvae, the risk of clams becoming contaminated by poisonous substances or pathogenic bacteria, and the risk of dying because of low temperature. All these things are important for clam cultivation, because they strongly affect the catch in the following year, the market value of the clams, and the total possible catch. Given these additions to the DEB scheme, it is better to refer to the kind of model demonstrated here as an **Individual Based Model**, or **IBM**. IBMs are frequently used when there is a need to predict what emerges from an ensemble of different individuals with different rates, ages, etc., because sub-models can be made for the different stages of the clam life cycle, and total quantities of clams in a given size range found by summing over the appropriate sub-models.

Another aspect of the conceptual model worth mentioning is the meaning of the boxes and the arrows in the diagram. At first glance it is intuitive. We immediately grasp the fact that phytoplankton is siphoned in by clams, which excrete some material in the form of faeces. But we also understand that the amount of clams or phytoplankton, or the temperature, are quantities that have a single value at any time, and that what we would like to predict is how they change with time because of the external forcing or their interactions. The interactions may be *fluxes*, e.g., the flux of nitrogen into the clam by siphoning phytoplankton, or may be *effects*, e.g., the effect of the temperature on the growth or mortality rate of clams. In our model the amounts, named state variables, are represented by the boxes, and the interactions (the fluxes and effects) are represented by annotated arrows. As we will see later, the next step in the formulation of a model consists in finding the appropriate mathematical formulae to quantify fluxes and effects (the arrows), so enabling prediction of how the state variables (the boxes) change with time.

Before moving on to that 'mathematicisation', let's finish with the conceptual modelling stage. The next step is to open up the 'clam fishery' box of Figure 4.2 in just the same way as we opened the 'clam' box in Figure 4.3. Figure 4.4 deals, largely, with economic information: the cost of harvesting clams in terms of labour cost, fuel cost, number of fishing boats and area covered; the cost of treatment if the site is polluted; the impact on the price by illegal fishery, etc. There are two new terms in this conceptual model: the propensity of customers to pay more for clams that are healthier to eat because they are, for certain, not polluted; and the effects of political deliberation and decision-making on waste treatment plants, on the amount of clams that may be

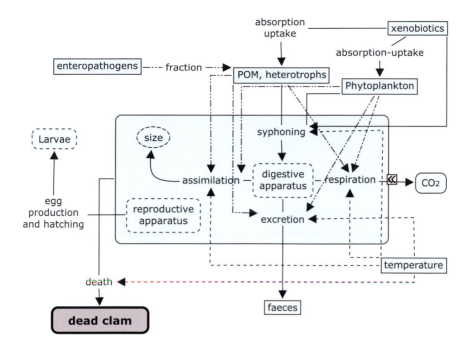

4.3 Conceptual Model for processes related to a single clam.

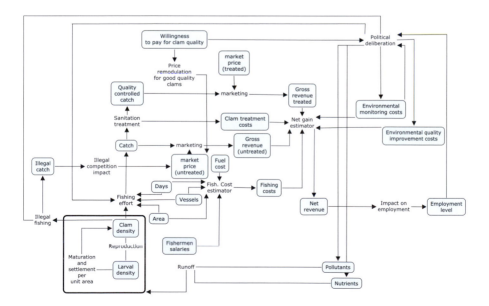

4.4 Conceptual Model for (mainly economic) processes related to the clam fishery.

caught, on the prevention of illegal fishery, even on the standards for clam treatment, etc. All these influence the net income of the fishermen, which determines the level of employment among the fisheries, which in turn feeds back on political decisions.

Hierarchy is an important property of systems. What we have been doing, in looking at clam processes in Figure 4.3 and economic processes in Figure 4.4, is zooming in on the contents of the conceptual model, moving down the hierarchy to examine the clam biology and clam fishery socio-economic sub-systems. Doing this as part of the process of System Design inevitably reveals more complexity than originally considered, and thus it often becomes apparent that the top level of the model needs to contain more, as is exemplified in Figure 4.5.

There are two key points to make about this revised conceptual model. The first concerns the boundary conditions: what lies outside the part of the system that we wish to model. Inside the system box (shown as a dashed-line rectangle) are components that interact with each other dynamically, which is to say that we expect change in one to affect another, and vice versa, during any simulation. External conditions act on the contents of the box, but are not affected by it: as exemplified by river inputs of nutrients, or by national legal constraints on shellfish cultivation.

The second concerns the interactions within the system box. These include feedback loops. That from 'Venetians' through 'sewage', 'contamination', 'clams', 'clam

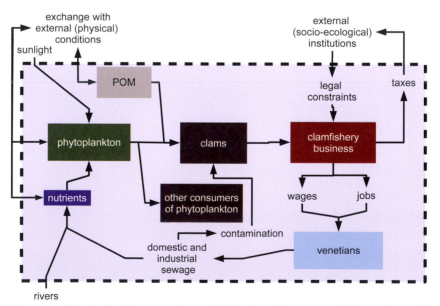

4.5 More complex top-level conceptual model. The dashed line marks the boundary of the virtual system, the part that is conceptualised in terms of dynamic interactions. The external conditions influence what is inside this system, but are not affected by it.

fishery' and back to 'Venetians' through 'wages' and 'jobs' is an example of a negative feedback loop. More sewage contamination will result in less income from the clam fishery, which will employ less people, and so, might help to stabilise the human population of the city. In many people's minds, sewage contamination is bad and population stabilisation is good – but the adjective 'negative' is not to be interpreted in terms of value, but only in the logical sense of Figure 3.4: when one stock or flux increases, the loop causes some other stock or flux to decrease. There is also a positive feedback loop, through sewage nutrients, phytoplankton and clams. This is 'positive' because the all three increase in parallel. That may be considered good (because there are more clams) or bad (because the increased phytoplankton might give rise to the harmful consequences that are symptoms of eutrophication).

These potential effects of increased phytoplankton biomass do not appear in the model. Nor do many other aspects of the real system. For example, tourism is not pictured in Figures 4.2 to 4.5, either as generator of wages and jobs, or a source of extra sewage. Nor have we considered other kinds of fisheries, for example of the crabs (moeche) which are a typical food for Venetians. These aspects are not considered to be of primary relevance to the *Issue* under consideration. The choice of what is important, and what is not, is an important part of a System Design process. In many cases it is based on the accumulated knowledge of the functioning of the system. In other cases it might be necessary to go through a trial and error process.

What makes a model tick?

A century ago, a *computer* was a person who did calculations, and there were proposals – for example – to predict the weather using a room full of clerks sitting at an array of desks, passing notes about the results of their calculation to their neighbours, and receiving equivalent notes from them. In the centre of the room a master clerk would ring a bell at the end of each completed cycle of calculations, to tell the clerks to start again, using their neighbours' results as new data. These proposals proved infeasible, but they correspond to what we nowadays do, using computers, under the control of a program that includes a cycle clock.

This cycle clock is part of the program, i.e. the software. Computers also contain hardware clocks, which provide an electronic pulse to synchronise the tiny flows of electrical current that are manipulated to correspond to the operations of arithmetic. Each tick of the cycle clock involves thousands, if not millions, of these pulses, and corresponds to a **timestep**. Timesteps are the little intervals of time into which a simulation is chopped to allow repeated calculation of changes in state variables that are interrelated. Timesteps need to be small so that values don't change much within them, otherwise interactions would be wrongly calculated.

Tick. The computer program causes electronic data to be passed in and out of the computer's processor in a way that solves the model's equations for a given set of conditions. The results, in most cases a new set of values of the model's state variables, are placed in memory. Tock. The process is repeated. Tick. And again. Tock. And again, until all the ticks and tocks add up to the passage of time that the simulation must encompass.

A modeller has a choice of two ways in which to make this happen. One is to make use of pre-existing software that allows the user to manipulate icons on the computer screen, with the software converting this manipulation into equations for calculating the state variables. The other is to write, explicitly, the instructions to the computer that constitute the model-equation-solving program. We will illustrate the second method first, starting with this short example of a program:

```
X = start_value;
for time = 0:timestep:365,  % in days
        X = do_the_model(time, timestep, X, conditions);
                        % 'conditions' is a set of values
    end
```

This an **algorithm**, a set of instructions (except for anything between '%' and the end of the line, which is a comment). The algorithm starts with the value of time set to zero and begins a set of calculations using the current values of the variables time, timestep, X and the set of values in the memory locations called 'conditions'. The instructions for the calculations are stored somewhere else, under the title 'do_the_model', and make up a sub-program or *function*. The result is a new value of X, which replaces the old value. Next, the value of time is set to its original value (zero) plus the value in 'timestep', and the sub-program 'do_the_model' is repeated, using the new value in the memory location called 'X'. And so on, until the value of time reaches 365 days, when the execution of the code is complete.

In fact, that code is the software clock; do_the_model contains the computer code for solving the model equations themselves, and it operates on the current value of the model state variable that is both symbolised by X and stored in a memory location named 'X'. Like this:

```
X = X * (1 + timestep * relative_rate_of_change_in_X);
```

Note that such algorithms do not follow the normal rules of arithmetic. Instead, the equals sign is an instruction to take that which has been calculated on the right-hand side of the equation, and put it into the memory location specified on the left-hand side. Because the latter refers to the same place as was used to supply the value of X on the right-hand side, this line of code has the effect of updating the value of the state variable. As we've already said, the value supplied for t i m e s t e p needs to be quite small (usually much less than 1 in whatever time units are used for the rates of change of the state variables) if this instruction is to give a reliable answer.

In addition, the sub-program needs a current value for r e l a t i v e _ r a t e _ o f _ c h a n g e _ i n _ X. Suppose X represents the current amount of phytoplankton in the Lagoon of Venice, quantified as the concentration of micro-algal chlorophyll in a cubic metre of seawater. Then the sub-program might include lines of code like this:

```
light_limited_growth_rate = photosynthetic_efficiency * ...
                (conditions.light - compensation_light);
nutrient_limited_growth_rate = maximum_growth_rate * ...
        conditions.nutrient / (half_saturation_conc +
            conditions.nutrient);
relative_rate_of_change_in_X = lowest_value_from ( ...
                light_limited_growth_rate,
                    nutrient_limited_growth_rate );
```

Note that the lowest level calculations, such as those of l i g h t _ l i m i t e d _ g r o w t h _ r a t e have to be done first, so that they can be combined in the line of code that calculates the r e l a t i v e _ r a t e _ o f _ c h a n g e _ i n _ X.

You may need to know a few more things about the computer programming language from which these examples are drawn. First, the set of characters that make up 'r e l a t i v e _ r a t e _ o f _ c h a n g e _ i n _ X' refer to a memory location, just as 'X' does: they form a label for a variable, and the value of this variable may either be supplied to the program in advance or calculated and re-calculated as the program runs. h a l f _ s a t u r a t i o n _ c o n c is an example of a variable supplied in advance, because it is supposed to be a constant property of the micro-algae that are simulated by the model. Such variables are often referred to as *parameters*.

Second, the label 'c o n d i t i o n s' has been defined, for this program, as referring to more than one memory location. Within this set, the specific location 'c o n d i - t i o n s . n u t r i e n t' contains a value for the current concentration of a nutrient such as nitrate. (Think of 'c o n d i t i o n s' as the name of a street, and of '. n u t r i - e n t' and '. l i g h t' as houses in that street.) Third, '*' is an instruction to multiply,

'. . .' shows that an instruction continues on another line, and ';' marks the end of the instruction.

Making models using pictures

As is well known, only a handful of people truly enjoy writing computer algorithms such as those we've exemplified above, and even fewer enjoy, or have the time and ability for, devising the formal mathematical equations corresponding to these instructions. It seems that many more people appreciate models drawn in pictures, such as those we have introduced as conceptual models of systems. Consequently, a number of software manufacturers have released computer programs that allow models to be constructed, and simulations made, by assembling icons on a visual display. SPICOSA modellers used the software **ExtendSim**® for this purpose, but in this section we will employ the simpler **STELLA**®. We'll use it to show how a sub-model can be constructed for phytoplankton.

Figure 4.6(a) is a sketch of a simplified micro-algal cell. It is in fact a conceptual model in its own right, because it shows the cell incorporating carbon dioxide into organic matter as a result of sunlight-driven photosynthesis, and then losing some of the organic matter as a result of respiration. In addition the cell takes up dissolved nutrients, including nitrates, from seawater. Part (b) shows how this might be modelled using STELLA®.

STELLA® is a useful teaching aid because its vocabulary – its set of available symbols – is very small. In early versions there were just four objects that could be displayed, moved around, and connected on screen. First, the rectangular box, which the STELLA® documentation refers to as a *stock*, and which we will equate with a system state variable. It is an amount of something, with a particular value at a particular time. In the preceding section, we said that the amount of phytoplankton, measured as chlorophyll, was an example of a state variable. In a clam–algal model, amount of phytoplankton is one of the things that we want to simulate, as food for the clams. However, constructing this iconic model is going to make us realise that we will need not one but two state variables referring to quantify phytoplankton. We will see why in a minute.

The second STELLA® object is a *flow*, represented by a double line with a tap symbol. Typically the flow object ends in an arrowhead attached to a stock, and it is useful to think of the double line as a pipeline, along which stuff flows. Clearly, the stuff in the pipeline has to be the same as the stuff in the stock box. Now, because photosynthesis is the process in which organic compounds are made from carbon

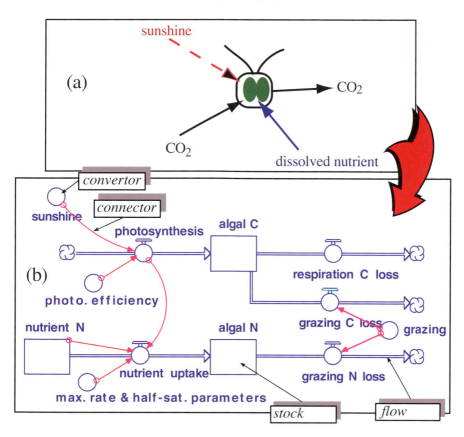

4.6 Using STELLA to model micro-algal growth processes.

dioxide and water, the photosynthesis pipeline must carry carbon, and it must be quantified as an amount of carbon in unit time. The box into which this is delivered, representing the stock of phytoplankton, must be quantified as an amount of organic carbon.

What about sunlight? Although it drives photosynthesis, it is not itself carbon. Therefore, in the present context, it is to be seen as information that acts on the rate of photosynthesis, and is described by the remaining pair of STELLA® objects. A circle, called a *convertor* is a general purpose repository, either for equations to convert one set of values into another, or for data, such as the average amount of sunshine on a typical day. The sunshine convertor is joined to the photosynthesis flow by a single line and arrowhead, called a *connector*.

The arrowhead on the connector shows where the information goes in, but doesn't specify exactly what STELLA® should do with it. To provide the software with sufficiently detailed instructions, the modeller has to open up the flow icon and type an equation into the resulting window, choosing from a list of connected objects to make something like this:

```
photosynthesis = photo._efficiency * sunshine
```

photo._efficiency is a value contained in another convertor connected to the flow icon for photosynthesis.

You have probably already noticed that the photosynthetic flow starts from a cloud icon. This tells us that STELLA® doesn't know, and probably doesn't need to know, what goes there. It might be that there should be another stock box, for dissolved carbon dioxide (mainly in the form of the bicarbonate ion) in seawater; if there were, STELLA® would automatically reduce the contents of that box by the amount of the flow of carbon into phytoplankton. If the source box doesn't exist, STELLA® in effect assumes that there is an infinite supply of material available to flow down the pipeline (controlled by the valve labelled 'photosynthesis') – in effect it is creating carbon out of thin air – or a cloud.

The use of such cloud symbols is a way of reminding modellers of the need to ensure that conservation laws are obeyed. A conservation law states that, for a particular substance, the total amount of that substance must be preserved by the calculation scheme within the model. Cloud symbols are 'exceptions that prove the rule': although the cloud that we've mentioned appears to create carbon from nothing, and there is another pair of clouds at which carbon disappears from the model, the carbon conservation rule requires that:

carbon into algal C *as photosynthesis*

   – carbon out of algal C *as respiration and grazing)*

      – change in algal C   *=*   *0*

We might, also, call this the 'Micawber principle', because it is a version of:

rate of change in state variable = total of input rates – total of output rates

and this is a general purpose equation that describes clam growth as well as micro-algal growth, and indeed also describes the economics of a business. If, as Charles Dickens' character Mr Micawber pointed out, income exceeds expenditure, all is well. If, on the other hand, expenditure exceeds income, a debtors' prison threatens. If a clam respires more organic matter than it captures in food, its biomass will decrease, and it will ultimately die. If an algal cell is unable to 'fix' enough carbon photosynthetically to balance respiratory losses of carbon, the cell will shrink and die. That is in reality, of course: but it is crucial that the model simulates such behaviour.

At first sight, grazing seems different. An algal cell either gets eaten, or it doesn't. However, the model is actually supposed to represent the dynamics of populations of micro-algal cells in the whole of the Venice lagoon. Each cell has a certain probability of getting eaten during a day, but so far as the population as a whole is concerned, this probability equates with a certain daily proportion lost to grazers.

Finally, there is another conservation law that states, in effect, that substances shouldn't be mixed. This is why there is a separate set of stocks and flows for nitrogen, which must be conserved as it is converted, from nitrate ions dissolved in seawater, into algal protein and nucleic acids. Of course, the nitrate concentration could be represented by a convertor, analogous to the sunshine convertor. But an important part of nutrient dynamics in seawater is the withdrawal of nitrates etc., and simulating this requires nitrate to be made a state variable.

How do we know this is right?

In the preceding sections, a magnifying glass has been moved over a few aspects of modelling, with the aim of giving the reader a feel for the processes by which conceptual models and simulation models are constructed. Although making conceptual models can be a participatory activity (as will be considered further in chapter 5), ensuring that the results of this activity are congruent with existing ecological and social knowledge requires considerable experience of these academic disciplines, and converting a conceptual model into a simulation model is a skilled technical task. How do we know that this has been done correctly, and that the results are right?

There are some technical answers to this question, but before exploring them, we want to broaden the debate: what does it mean to claim that model results are correct – especially in cases when the model is simulating an imagined future? Are we, firstly, arguing that the conceptual model is correct – i.e., that it does not contradict what is known in general about social-ecological systems and about the studied system in particular? Or, rather more strongly, that it is the best of all the possible conceptualisations of the system that are relevant to the Issue? Are we, secondly, claiming that the algorithms of the simulation model instruct the computer's numerical processor to carry out operations that result in the modelled state variables changing in the same way as they would in the real system? And, thirdly, are we claiming that we've supplied the simulation program with sufficiently reliable data about, for example, sunshine, so that an actual set of simulation results would, if compared, be found to be in good agreement with observations of the corresponding reality?

In some cases it is possible to test whether model results are correct, by comparing them with known answers. Sometimes these answers can be obtained independently from scientific theory, but the most satisfying tests are those involving confrontation with the real world. An example is shown in Figure 4.7. These result from a test of a phytoplankton model, programmed in ExtendSim®, against data for a Scottish fjord that has been observed during four decades.

The model state variables included the concentration of phytoplankton chlorophyll in the superficial waters of the fjord. The software was asked to simulate changes in this variable over a calendar year, outputting one value every day so that chlorophyll could be graphed against the day of the year. In order to simulate these changes, the modellers had to provide the program with data on relevant conditions – including daily values of sunshine and the concentration of chlorophyll in the water outside the fjord, with which the fjord contents were exchanging. Here's a final algorithm, written as computer code:

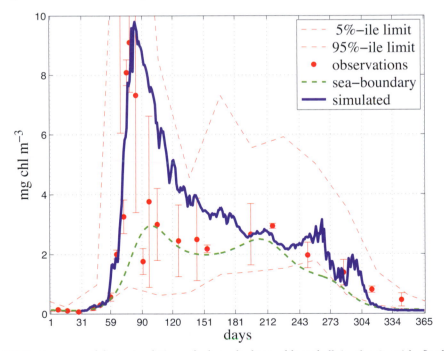

4.7 Testing a model. A simulation of phytoplankton chlorophyll in the Scottish fjord, Loch Creran, for 1975. The dashed red lines give upper and lower limits to amounts of phytoplankton chlorophyll observed in the fjord between 1970 and 1976 (see Tett and Wallis, 1978). The red-filled circles are the means of observations (with lines showing ranges) made in 1975.

```
final_relative_rate_of_change_in_X =
                    within_fjord_rate_of_change_in_X
                         + exchange_rate * (X_outside - X);
```

The `within_fjord_rate_of_change_in_X` is the net rate of change calculated as a result of all relevant processes within the modelled virtual system. It is the term `relative_rate_of_change_in_X` corrected for losses, such as those due to grazing by planktonic animals and by seabed animals such as mussels and clams. The parameter `exchange_rate` is the average proportion of water inside the fjord that is replaced, each day, by water from the external sea. `X_outside` refers to the concentration of chlorophyll in this water, and is an example of a boundary condition. Figure 4.8 shows an implementation of the algorithm in the symbols of ExtendSim®.

A general difficulty in modelling is to get good data about the boundary conditions. They are, by definition, likely to be somewhat distant from the site of interest. For example, in the case of the Scottish fjord, there were many data from the fjord itself for most of the years from 1972 through 1981. Data about sunshine could be obtained from the national meteorological office. But the sea outside the loch had been sampled for chlorophyll, mainly in 1970, and then not again until 2001.

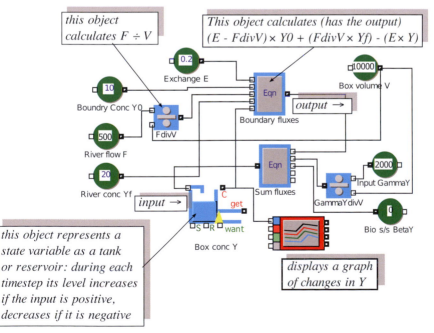

4.8 Example of an ExtendSim® block used in the simulation model that output the data plotted in Figure 4.7.

Nutrient data were equally sparse. So the modellers constructed a climatology, a set of data that described how on average chlorophyll and nutrients changed at the fjord's sea-boundary during a typical year. For comparison, all observations from the fjord itself were used to construct a slightly different sort of climatology: one showing, not the average pattern of seasonal change in chlorophyll concentration, but the limits within which chlorophyll varied. We'll call this an *envelope*.

The simulation of the state variable, chlorophyll, shown in Figure 4.7, was calculated for the year 1978 – meaning that sunshine and other meteorological data used by the model were the values measured on each day in that year – but was provided with climatological boundary conditions. Therefore, two tests of the simulation were possible. The first was the weak test of checking that the simulation lay within the fjord's climatological chlorophyll envelope. The second was the stronger test of seeing how well the simulation agreed with chlorophyll concentrations observed in the fjord during 1978. In this case it was found that simulated values differed from observed concentrations by an average of 30% of the observed values. To put it another way, the model was able to explain about 70% of the observed variation in chlorophyll in the fjord. Given the imprecise boundary conditions, that was a satisfactory result.

Can we do it?

The tasks that have been illustrated above can be demanding of time and technical skills. Furthermore, scientists, by nature and training, wish to pursue understanding, and understanding a social-ecosystem is a big challenge. A SAF application, however, needs to deliver results in a timely fashion, meaning when needed by the stakeholders, rather than when fully satisfying to scientists. Thus the process of designing and building a model needs to be managed for delivery rather than new scientific results, by drawing as far as possible on existing knowledge and building on existing models. Icon-based modelling software allows new model parts to be assembled quickly. There is a narrow path to be steered, between excessive simplification of the virtual machine, which may fail to capture key relevant processes in the real system, and the inclusion of too much complexity, which will cause delay and expense, and will not automatically result in better simulations, because more complex models need more data to be supplied for parameter values and boundary conditions.

Given experience, then, and a focus on models that are just as complex as they need to be, but no more, models *can* be built and run in the time available. Of course, the necessary complexity should include all socio-economic as well as ecological processes. This stipulation returns us to the question asked in chapter 2: how much of the social system should be quantified and included in the model? This is a question about simulation models, because it is no more difficult to include a social component in a conceptual model than it is to include an ecological component. The parts of the conceptual models in Figures 4.4 and 4.5 that deal with external social constraints can be expanded into an *institutional map* showing the several organisations of the Italian state and local government, and the fishermens' collectives that need to be considered in relation to the Issue (Figure 4.9). The difficulty over the social component lies not in conceptualising and mapping, but in quantifying and simulating it.

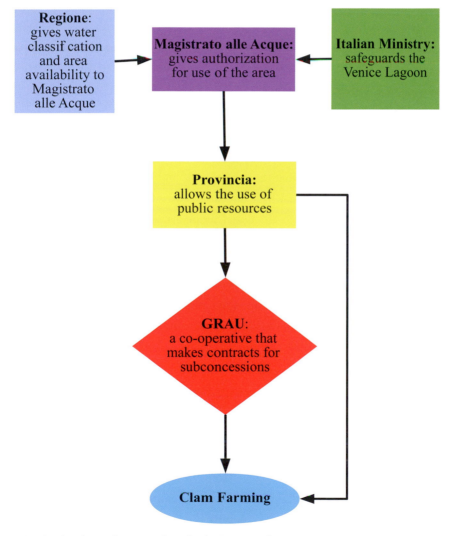

4.9 Institutional map for aquaculture in the Lagoon of Venice.

Within the SPICOSA project, different study site application teams reached different conclusions about this. Some preferred to treat the social part as relevant to identifying stakeholders and defining the Issue, and then assessing the results of the model, but as essentially qualitative and so unfit for description by algorithms that implemented mathematical statements. Other teams used economic concepts to bridge the gap between ecosystem and social system. For example, the clam fishery business in Figure 4.5 converts the lagoon's biological productivity into a cash flow and thus wages and jobs, which, in classical economic theory, helps satisfy the well-being needs of Venetians. Another example is provided by the SPICOSA study of Himmerfjärden, south of Stockholm. The Issue at this study site concerned the risk of eutrophication, the stimulation of excessive algal growth by human-released nutrients. Swedish people value clear water, and studies have shown they are willing to pay for this through taxes or charges. Thus a simulated increase in water transparency, through reduction in the amount of phytoplankton in the water, was linked to a simulated income from 'willingness-to-pay' for clear water (Franzén *et al.*, 2011*)*.

Using a Model

We started this chapter with a brief account of modelling storm surges in the Lagoon of Venice, making the point that the atmospheric and oceanographic models used to predict periods of high water have one simple use: they aim to forecast the near future with sufficient reliability for Venice's shopkeepers, residents and visitors to adapt their behaviour. SAF models have a more complex use, in comparing the likely outcomes from different scenarios.

As explained in earlier chapters, scenarios can be thought of as quantified stories about possible futures. In the case of the Venice lagoon study, they concerned alternatives for management of nutrient loading on the watershed draining into the lagoon, alternatives for climate change, a range of market prices for harvested clams, and alternatives for clam cultivation practice. Only the last are under the control of the clam growers and the authorities who regulate the concessions, the others making up a range of possible changes in boundary conditions – things imposed on the modelled system from outside. Under present-day conditions, the clam bio-economic model showed that the best cultivation scenarios depended on the market price for the clams. If small clams fetched a good market price, then the greatest revenue was obtained by using a short cultivation period. This, however, proved the most damaging to the sediment. The most sustainable outcome was obtained with a longer cultivation period, leading to larger clams and less overall re-suspension of sediment.

We end this chapter and its account of how to apply a Systems Approach Framework, at the end of the steps that make a heavy demand on the technical skills of system designers and the formulators and users of simulation models. Once the modellers have completed their work, the resulting simulations need to be interpreted and presented to stakeholders for deliberation, leading, ideally, to a decision either by them or by public officials to choose whichever scenario, and hence the management scheme, that both optimises sustainability and protects their interests. The topic of society's use of model results is, however, the subject of the next chapter.

EndNotes

Works that provide guidance for and insights into the use of system simulation models include: Bellinger (2004); EPA (2009); Fennel and Neumann (2004); Morecroft (2007); Odum & Odum (2000), Soetaert & Herman (2009).

[Problems in the Lagoon of Venice] The cause of the *acqua alta* is this. Normally air pressure is high over the Mediterranean region, and atmospheric depressions traverse the Atlantic at higher latitudes, bringing wind and rain to northern Europe. In winter, however, the tracks of some depressions run south of the Alps. As these low-pressure regions pass over northern Italy, higher atmospheric pressure above the Mediterranean itself pushes water into the Adriatic. This 'atmospheric tide' adds to the small astronomical tides, and is intensified because the Adriatic is a long narrow sea.

[Designing a system and formulating a conceptual model] Examples from the Lagoon of Venice are based in part from Melaku Canu *et al.* (2011) and internal SPICOSA reports generated by the study site team, but freely adapted for our purposes here. Figures 4.3 and 4.4 were drawn with Cmap software (cmap.ihmc.us). Grimm & Railsback (2005) is a textbook of individual-based modelling. *See* Bacher & Gangnery (2006) for an example of both DEB and IBM.

[What makes a model tick?] This section attempts to bridge the gap between the mathematical world and the world inside computers. The code examples in this section are written in a computer language called Matlab, part of a software package that includes an extensive library of mathematical routines, including those for numerical integration, manipulations of matrices or arrays as well as single variables, and for graphical output. The package is available from the MathWorks Inc.

(http://www.mathworks.com/). The algorithm that starts the explanation of do_ the_model uses the simplest possible (Euler) approximation for the (numerical) solution of an integral equation. Given the mapping,

$$X \rightarrow X + \int\limits^{\Delta t} \frac{dX}{dt} dt \quad ,$$

the (forward) Euler approximation is:

$$X \rightarrow X \cdot (1 + \Delta t \cdot \frac{dX}{dt} \cdot \frac{1}{dX}) \quad .$$

For details of more reliable and precise numerical methods, see, for example, Press *et al.* (1989).

[Models in pictures] The emphasis in this account is on systems of state variables in which change can be described by Newtonian or Leibnitzian differential equations. For example, the rate of change of phytoplankton biomass X might be described by:

$$\frac{dX}{dt} = (\mu - g) \cdot X \quad ,$$

where μ refers to relative intrinsic growth rate (the result of biomass increase and cellular multiplication) and g to the relative rate of loss due to grazing by larger animals. Software such as STELLA® (www.iseesystems.com) and ExtendSim® (www.extendsim.com) automatically sets up the algorithms for numerical integration of the rate-of-change equations, although the user would be wise to check the results of these algorithms for accuracy using test cases with known answers. STELLA® is more limited in its capacities for ecosystem modelling, but we have found it very useful for introducing modelling to university undergraduates in the life sciences. As mentioned, it has a restricted symbol set. Some writers (e.g. Heemskerk *et al.*, 2003) advocate using a wider range of symbols in conceptual models.

[How do we know that it is right?] An overview of the philosophical basis of scientific modelling is given by Frigg & Hartmann (2006). Alexandrov *et al.* (2011) discuss how scientific journal reviewers should view claims about model adequacy, verifiability, reliability, and legitimacy. Refsgaard & Henriksen (2004); and Rykiel (1996) deal with validation. The model validation work in Loch Creran was part of the SPICOSA study at site 7, and is based on Tett *et al.* (2011). *See also* Laurent *et al.* (2006) and Portilla *et al.* (2009). The error figures mentioned in the text are root-mean-square errors, the square root of the mean of squared differences (along the vertical axis) between simulated and observed values. Also useful is r^2, the proportion of observed variance that is accounted for by the simulated values.

CHAPTER 5

Bridging the gap between science and society

Anne Mette

Introduction – the need for communication

Imagine a fisherman in a quiet and remote area somewhere along a southern European coast. He knows from his 36 years of experience that if the number of big vessels in his little fishing area is increased, his catches will get less. He therefore decides to go to a public hearing where some commissioned research results about the local fisheries are to be presented. But he does not have good skills in mathematics, and, since high school, nobody forced him to think in three dimensions. As a consequence, he does not understand any of the diagrams shown in this meeting. Nevertheless, he gets the key point: based on these cryptic data, he sees that the plan for more big vessels will get the go-ahead, damaging his livelihood, and, as a bonus for the local economy, three new hotels will be built, endangering his little cottage by the shore. Meanwhile, the regional government praises itself for its participatory approach to coastal zone management.

This is, perhaps, an extreme example, but it makes two points. The first is about a failure in communicating scientific data *to* an audience. The second concerns a failure in communicating *with* that audience. An opportunity for dialogue has been missed. This chapter is about the two forms of communication: how to communicate scientific information wisely *to* stakeholders and public officers; how to ensure communication *with* and *amongst* these groups, so that the outcomes of participatory processes are seen as legitimate and so that stakeholders obtain **ownership** of the processes of sustainable development.

Natural systems can be said to communicate with each other, through the flows of mass, energy and information along the pathways of interactions between system components. But this is not communication in human speech; it must be read from the signals sent by ecosystems, and, as illustrated in chapter 2, humans sometimes only notice the ecosystem's way of communicating when it is too late for action. We do not want to claim that scientists speak *for* the ecosystems. But we do claim that they can enter a dialogue with coastal citizens to exchange knowledge *about* these natural systems and to understand better their links with the social systems in the coastal zone.

Whilst preparing this book we have asked ourselves why 'integrated coastal zone management' has proven so difficult, despite many people trying it (Shipman & Stojanovic, 2007). Perhaps it is because of the tough task of building a communicative capacity between science, governance or policy-making, and civil society; and amongst economics, ecology and social science. The methods and findings of the natural, or 'hard', sciences are not always easy to adapt or extend to the social system. A mussel farmer might understand a suggested management plan very differently from the way it was conceived by the scientists who did the modelling. If a wild salmon were able to express its opinions, it might complain about the obstacles it has to pass during its travels. And neither salmon nor mussel farmer is likely to be fully aware of the complex interaction between local, non-local, and social-ecological dynamics, when policy is made. For the farmer, this knowledge-gap could lead to his rejecting the policy-maker, to voting for someone else at the next election, and to joining a producers' organisation to defend his stakes. Whereas for the salmon, unsustainable management plans might imply losing either its life or that of the following generations.

You have, so far, learned about the history and importance of our planet's coastal zones, about institutional and governance rules relevant to social-ecological problems, about systems thinking, and about the virtual world of modelling. Half of the planet's human population lives on the 10% of land that is considered as the 'coastal zone'. How can they, and their interests, be taken into account in understanding the coast's 'breathtaking complexity', and in sustaining coastal zone systems, as promised by the title of this book?

Our argument is that a Systems Approach can provide an intellectual framework for this understanding, and can help to open a space for communication and dialogue. It offers methods to bridge the gap between science and society. Hence, this chapter will bring together the pieces of the puzzle, and will complete the description of methods useful for applying a Systems Approach Framework, or SAF, to sites and problems in the coastal zone. These methods were refined and tested in the European research project SPICOSA, which responded to the need for an improved science–policy and science–society interface to manage our coastal systems the most sustainable way (Hopkins *et al.*, 2011).

This chapter will explain the importance of dialogue, transdisciplinarity, participation and deliberation to pave the way for successful science–policy integration in a Systems Approach Framework. It will also give a brief account of the five steps of a SAF application (Figure 1.5). These are: the identification of a policy issue in the Issue Identification Step; the design of the system that corresponds to this issue and its various dimensions in the Design Step; the mathematical formulation and assessment of the socio-economic and social-ecological dimensions in the Formulation and Appraisal Steps; and the visualisation and translation of scientific findings for the work with scenarios in the Output Step.

What do you need for a SAF application?

This section deals with those aspects of communications and stakeholder engagement theory that ought to be taken on board before launching an application of the Systems Approach Framework.

Acknowledging Cultural Dimensions

Imagine asking a natural scientist to consider cultural differences throughout the European Union when starting a dialogue with stakeholders and environmental managers. The scientist might say, 'Surely, science is the same everywhere', or, 'I have never been to Greece, so how should I know what they do there?'

In the terms introduced in chapter 3, the rules of the physical world 1 are indeed the same everywhere, but the rules of world 3 are less consistent. Humans can remake them locally, either intentionally or accidentally. SPICOSA tested the SAF at 18 sites across Europe and its near neighbours (Figure 1.4), and found many differences in 'best practice' in convening meetings, in what was considered proper in deliberating and entering dialogues, in obtaining data for integrated virtual models, and in attitudes to the environment and the characteristics of traditional ecological knowledge. For example, in the west Highlands of Scotland, there is much emphasis on oral exchange of information, whereas in northern German Baltic coasts, people prefer to prepare by reading written briefings. In some regions, stakeholders were happy to look calmly at mathematical formulations, while in others, scientists found themselves confronted with furious fishermen and had to be rescued by the police (Figure 5.1).

It is unsurprising to find differences between what people feel free to say at meetings in Scandinavian social democracies and in countries on the other side of the

5.1 Newspaper articles about people charged for illegal fishery and a public meeting which ended in a fight. *Il Gazzettino*, Venezia, 9 May 2008.

Baltic that were, until recently, part of 'actually existing socialism', even if many of them share the inheritance of Martin Luther. Our imagined Nordic scientist doesn't need to travel from the Himmer fjord in Sweden to the Salonic Gulf in Greece to appreciate that such differences and such common ground might exist. The appreciation comes from the realisation that applying a SAF is also a socio-cultural experiment and learning process and more than just the making and using of a mathematical model. And we found that scientists carrying out the SAF applications had learnt not only from stakeholders and policy-makers, but also from each other's research disciplines.

This is making the point that there are several cultural dimensions involved in a SAF application: the regional dimension; a dimension relating to level of formal education of stakeholders; and the dimension on which are spread out the different scientific disciplines and technical skills that need to be brought to bear. It was anticipated that Dr X, lightly built and wearing spectacles, might feel nervous when meeting with the burly fishing skipper Captain Y, or the forceful Madame Z, former party secretary. What was less anticipated was that Professor W and Dr X might fail to communicate when the first understood 'integration' as the political process in which Europe is engaging, whereas the second wanted to know whether to use a Runga-Kutta or Euler numerical scheme to solve equations in model state variables.

A SAF application needs to communicate with and engage stakeholders, which means taking account of their points of view. And it also needs to be transdisciplinary: it needs to describe a problem from the 'problem's point of view' rather than the separate points of view of several disciplines. Our salmon and our puffin will find themselves in the mathematical model together with algae, with phosphorus or other elements, and with species that might endanger (or not) their well-being. Our *Homo sapiens littoralis et societatis*, however, will also play a central role in this process – he and she will be part of the transdisciplinary team: scientists in dialogue with policy-makers and stakeholders.

Dialogue
Coastal citizens are at the moment not fully prepared for the rapid changes in our coastal zones. Why write something so unhelpful? Because it leads us to explain several ways and tools of communication needed for sustaining the coastal zones. The first of these is the communication of complexity.

Jumping back in time, almost five centuries ago the German Martin Luther – the initiator of the Protestant Reformation – gives us an example for the importance of communicating complicated things. It would have been the contemporary procedure to translate the Bible from Greek and Hebrew to Latin. Luther, however, did not do so, but translated it into popular German. By doing so, he made it possible for the German people to understand the Bible.

A similar challenge faces today's coastal zone management: that of communicating complex issues and complicated scientific results to stakeholders, policy-makers and environmental managers. As in Luther's case, the first step is to listen to the language in which people speak amongst themselves. The next step is to go beyond 'simply picking the people up where they are', as used to be said of Luther. It is necessary to offer new thought patterns and mindsets, and the possibility of grasping complexity and novel perspectives.

Before being actively engaged in the process of science–policy integration, stakeholders do not necessarily have the skills for ongoing or rapid cultural transformation processes (Borner, 2011), meaning the ability to adapt to forces of different origins (ecological, political, economic, technological, to name only a few). The use of scenarios and conceptual models has been found to be a good way of bringing together the different viewpoints and lifeworlds of stakeholder and scientists, and can start and support the process of mutual engagement.

Communicating complexity is related to intensive (social!) dialogue: to value-driven exchange of opinions, consultation, coaching, and deliberating. We argue that these integrative communication processes are necessary complements to the functional and technical communication of scientific knowledge for steering the decision-making processes towards a sustainable development.

We are now – maybe without yet being aware of it – entering Jürgen Habermas' theoretical realm of the lifeworld ('Lebenswelt' in German) of communicative action. Habermas speaks about cooperative action that is undertaken by individuals and based on deliberation. This indicates that actors (in our case, stakeholders, policy-makers, and scientists) seek to reach a common understanding and to act based on reasoned argument and deliberation rather than only pursuing their own goals (Habermas, 1981).

Habermas' former student Niklas Luhmann also stresses the importance of communication, in fact, describing and explaining society as a complex system of communications. Habermas and Luhmann have disagreed over their interpretations (Leydesdorff, 2000), but Luhmann is useful to us because of his interest in information coming to society from what surrounds it – and part of this is the natural environment. Thus his most relevant book is *Ecological Communication* (Luhmann, 1989). For him, society consists of nothing but communication and hence, he goes as far as claiming that something that is not communicated does not exist in society. Think about this literally, because it has consequences also for ecological problems: if no more wild salmon cruised rivers and seas, if a pair of puffins were the last of their species, or if the German Oder Estuary came close to witnessing a 'tipping point' – these things would have no meaning if nobody knew: they only become problems once they have been communicated (Borner & Bittencourt, 2003). Perhaps a brown bear would be a little hungrier if there were no salmon runs, but it would not know that a part of its species' usual diet had vanished.

Luhmann's theory implies that sustainability requires, firstly, intercommunication amongst distinct social systems such as politics, law, science, economy, and civil society. Secondly, there must be communication of the existence and complexity of coastal zone problems. And, thirdly, there must be communication about the complexity of the communication processes themselves. *Communicating sustainability* is therefore a process of understanding, dialogue, and deliberation, and is not only about exchanging information but also about mutually discovering the way towards sustainable development (Michelsen & Godemann, 2007).

Change management includes the communication of information about transition processes and about the measures that are needed in order to keep up with, or adapt to, the changes. These measures can include pre-emptive adaptation of social structures, but it is necessary to understand that simply presenting information about change is insufficient: those likely to be affected – the stakeholders, with their stake in what is going to change – have to accept the information and the proposed restructuring, or what they think will change for the worse (Klewes & Langen, 2008).

Science communication is the transfer of scientific findings, including the reduction of these findings to what is most relevant to, and most comprehensible by, the respective target groups. However, such simplification has to be done carefully. The aim is to communicate neutrally, without bias in favour of particular decisions by these groups. Such even-handedness is one of the key skills needed for engagement, and is hard to maintain. Communicators need to know their own mind-sets and potential biases, and should try to correct for these. Target groups should be aware that no scientist can ever be completely neutral. However, the bias should only come to show when the scientist puts on a stakeholder's hat – for scientists can also have stakes – and not during the work for communicating science.

Scientific discourse, referring to equations, fractions, arithmetic, chemical elements, or the Latin names for different types of algae or mussels, etc., is usually the wrong language for a presentation to the public in general, or, in most cases, to specific target groups, such as stakeholders, managers or policy-makers. A fisherman will know the names of the fish he catches; a mussel farmer will know the best months of the year to harvest the mussels. Both might know what impact an increased sewage discharge has on mussels and fish. But neither has a reason to know in detail about the biochemical processes that impact the quantity of their harvest and thus affect their economic well-being. Nor will they know which equations are needed to construct a model to simulate these processes. Building a model and getting it to run is a sophisticated technique for experts. Stakeholders listen and translate within their **codes** – their interpretative frameworks – and if they hear something different to their present understanding, they will tend to ignore it (Luhmann, 1989). In which case, scientific knowledge will not make it into the heads of members of civil society, and knowledge about the possible futures of the coastal zones will remain locked inside scenarios.

Risk assessment looks at (transformation) processes and appraises whether these are risks or hazards. Risks are manageable, whereas hazards are not manageable. *Risk communication* is the interpretation and social communication of this assessment. For example, a burning oil platform is a hazard to its crew, and any resulting oil spill is a hazard to marine ecosystems. But as drilling companies like to point out, an oil platform in normal use is a low risk to humans and environment, because proper operation procedures minimise the probability that there will be a fire. Another example: the Wadden Sea will gradually be submerged by sea-level rise. It cannot retreat and regenerate, because of dykes behind. Nature lovers (enjoying its cultural services) and ecologists (speaking for nature) will view sea-level rise as an existential hazard and a possible catastrophe for the Sea. For the climate-change denier there is no risk, and for the hotel owner who took out insurance before scientists began to estimate the consequences of melting ice there may be no hazard, but only the inconvenience of rebuilding on a new site.

All these varieties of communication may be needed in an active and participatory dialogue about sustainability, a mode of approach that is beginning to replace the earlier mode involving communication of environmental risks to a passive audience. The older mode aimed to communicate the need for sustainability by talking about the complexity of nature. However, there is increasing evidence that 'talking at' an audience about sustainability fails to bring about changes in outlook; indeed, may often prove counterproductive (Franz-Balsen & Heinrichs, 2007). The new mode of *sustainability communication* adds participation and **social learning** (Michelson & Godemann, 2007) to the communication of science, sustainability and risk. That is the reason why this subsection was titled 'Dialogue' although it has dealt mainly with communication. In the next subsection we look at the other participants in the dialogue, the stakeholders.

Working with stakeholders

General Charles de Gaulle is reputed to have said that 'politics is too serious a matter to be left to the politicians'. Politics is the process by which people make collective decisions, and the Greek root meant either 'citizen' or 'affairs of the city'. As individuals, stakeholders are sometimes members of loose subgroups of citizens. In chapter 1, a stakeholder was introduced as someone who has a moral interest in a social-ecological problem, because they cause it, suffer or benefit from it, are concerned about it, or are responsible for managing it. We argued that the most efficient, fair, and sustainable, solutions are most likely to be reached when stakeholders take part in identifying the problem, the building of scenarios, and in appraising the results of the scientific simulations and findings.

In a nutshell: the prospects of bridging the gap between science and policy is greatest when the stakeholder's interest's relevance is acknowledged by scientists and policy-makers, and used to steer the SAF application towards greater relevance.

Scientists move on to new problems when their research funding runs out. Public officials, whether making implementation decisions or policy, move on to new problems when they have found a solution. The term of a politician might end at the next election, after only a few years. Stakeholders are the people who must implement policy decisions, can support them or in various ways ignore or subvert them, and in any case must live with them (whether they agree with them or not). Depending on their social experience of decision-making, stakeholders may behave actively or passively, but it is desirable to see them as dynamic practitioners, contributing both traditional ecological knowledge, and information about stakeholders' perceptions of their own needs, to the analysis of the problem.

So it seems to be no more than common sense to involve people in debating matters that impact on their lives and livelihoods. Our argument goes further, however, drawing again on the work of Jürgen Habermas (1981), who claims that the ability to communicate with each other is a key result of human evolution and is thus built into our natures. By talking sufficiently and seriously together, people can sometimes reach a consensus about the natures of, and the solutions to, problems that affect their interests. They have the potential to be more creative and dynamic in implementation processes than scientists, environment managers, or policy-makers can be within their existing institutional and governance frameworks. This is not always the case, but even when there is no agreement about what to do, shared deliberation of different views can lead at least to increased empathy and understanding (of an individual's own interests as well as those of others), and so to greater trust amongst users of ecosystem services even when there is competition for those services. Thus the rationality that emerges from deliberation will often make more sense than the sum of individual rationalities – or, even, than a political decision by a majority.

These ideals, however, can run into the quicksand of reality. Scientists and public officials, paid to spend time on engagement, may think, quite rationally, that a stakeholder's interest will automatically lead them to come to meetings or take part in electronic debate. But stakeholders often lead busy lives, and know their own interests quite well. They may have had bad experience of consultation in the past, or they

may wonder why they are giving their time for free to some scientific research programme. It helps greatly if there is already local experience of stakeholder involvement, or existing forums that can be used for the present purpose. If not, trust must
be built from the ground. Offering lunch, for example, not only compensates stakeholders for lost time, it begins to demonstrate that they are valued, and it draws on
traditions of hospitality that are found in most societies.

Stakeholders often fall into groups with conflicting interests. Some of these
groups might be larger, or better organised, than others. They might choose to pay
for a skilled and articulate representative to take part on their joint behalves. A stakeholder might be a legal person but in actual fact a company, and send their lawyer as
a representative. Others might attend as individuals, motivated by a strong concern
for their own interests, but lacking confidence in their articulacy in public. It is for
all these reasons that there is a need for specialised procedures, or tools, to aid participation and deliberation.

Participation tools

Although we write about stakeholders, the term is better understood as referring to
a relationship between a person and an issue – i.e. about stakeholding. The above-
mentioned theory of communicative rationality was set out by Habermas (1981),
and is summarised by Finlayson (2005). Habermas claims that speech acts have a
goal of mutual understanding. We interpret this to mean that such understanding
was a benefit that selected for an inheritable ability to speak and understand speech
during human evolution. Habermas contends that in the modern world, this more
humane rationality – which belongs to what he calls the *lifeworld* – may be overwhelmed by the instrumental rationality of what he calls the system of large-scale
institutions and organisations. Some of our methods and tools are intended to help
bridge the gap between, on the one hand, the lifeworld – the 'real world' in which the
stakeholders live – and on the other hand, the large-scale institutions of governance
that are needed in large and complex societies, and the scientific knowledge needed
for sustainable development.

For realising this integration – for entering a dialogue and deliberation process
among the heterogeneous stakeholder groups and the scientists – the SAF draws
on participatory tools and methods. The tools needed are tools that foster not only
the participation of stakeholders but also help to share an understanding of their
worldviews (in their lifeworlds), stakes, knowledge (and knowledge gaps) and consequently the language codes with which they communicate. The tools should en-

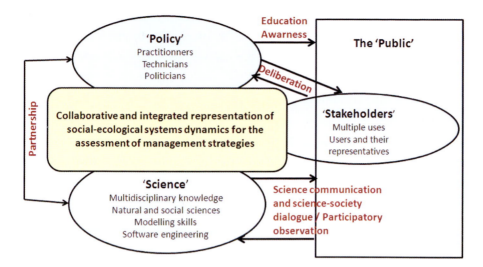

5.2 Diagram about the collaborative interfaces in a SAF to illustrate the communication flows and collaboration processes within a SAF.

courage non-scientists to explore and specify a selected issue, in all its dimensions, so that their views and knowledge can feed into a conceptual model and help to formulate a mathematical model. Using these tools, stakeholders can achieve owner-ship of the process, which could then also be called an empowerment process. Fur-thermore, participatory tools contribute to ensuring the relevance of an integrated assessment by making it more robust in the social dimension and by improving the quality and relevance from society's point of view. Thus they help to guarantee the legitimacy of the process (Pereira & Brilhante, 2005). The flows between science, policy and stakeholders in a SAF are visualized in Figure 5.2.

Deliberation and supporting tools

Coastal zone issues are complex social-ecological problems. We might dream of the ideal deliberative democracy, in which stakeholders drive both science and policy-making in a way that reflects both the complex societal context and the need for sustainability. But whatever the organisation of a polity, research stresses that 'delib-eration is the most effective form of science and policy integration to increase social learning and to achieve more useful and innovative approaches to environmental management' (Mette *et al.*, 2011). This sub-section aims to provide an insight into deliberative elements and supporting tools for coastal zone management.

Dialogue, discourse, discussion, deliberation – lots of Ds! Deliberation incor-porates the three others. It is a process in which individuals and organisations are open to analysing and changing their preferences through a process of persuasion (but not manipulation, deception or force) by other participants. It takes place in an open process of dialogue, discourse, discussion, and exchange of knowledge and ideas (Mette *et al.*, 2011).

Decision support tools or **decision support systems** refer to a wide range of tools which support decision analysis and decision-making and participatory processes. It is a crucial requirement that such tools, in addition to their primary functions, are themselves easy to understand, so that their use can foster communication and deliberation – an informed debate – amongst stakeholders and policy-makers with different backgrounds.

Although models are technically challenging objects, they can also serve as support tools. As described in chapter 4, a complete simulation model is likely to comprise a database, computer code to solve the equations of the mathematical models representing the ecological and socio-economic systems, and an interface for visualising the results of simulations. As explained in chapter 3, its core purpose is to show 'possible futures' in the form of simulations driven by scenarios. However, by designing the interface to be comprehensible by non-specialists, a model can be made to provide a wider range of insights. Most people can 'read' conceptual models, if suitably presented, and it turns out that including some or all of the conceptual model in the visualisation interface gives non-specialists a much greater understanding of what is going on in that otherwise mysterious 'black box'. If stakeholders have helped design the conceptual model, then our evidence from SPICOSA is that, not only can they 'read' the model and its results more easily, but also that they attach more legitimacy to these results: they have already 'bought in' to the modelling process and its outcome.

The second key set of tools useful in a SAF application are those used for, or in relation to, deliberation, and, hence, given the acronym DST (Deliberation Support Tools). Sometimes they lead to decision-making, but that is not their main purpose, which is as a vehicle to facilitate dialogue, knowledge transfer, and the exploration of that knowledge. Secret voting, the essential tool of a modern democracy, is not a deliberation support tool, because it excludes the possibility of people explaining why they voted in a certain way, justifying that vote, and listening to others' justifications. A **forum** is such a tool. Of old it was a public place, a market both for goods and information, as well as, sometimes, a place to vote. Modern forums are, typically, meetings for discussion and deliberation, and they are usually convened, managed, policed, moderated or facilitated in some organised way.

Stakeholder Name / Category ....

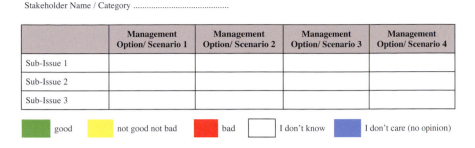

	Management Option/ Scenario 1	Management Option/ Scenario 2	Management Option/ Scenario 3	Management Option/ Scenario 4
Sub-Issue 1				
Sub-Issue 2				
Sub-Issue 3				

🟩 good   🟨 not good not bad   🟥 bad   ⬜ I don't know   🟦 I don't care (no opinion)

5.3 Deliberation sheet for a deliberation matrix.

One tool to bring together different categories of stakeholders with rights and interests of various strengths, and policy-makers with various degrees of influence and power, is by means of a **deliberation matrix**. Such matrices can be used both at the start and at the end of a SAF application. Figure 5.3 shows a simple start-up example, a table in which the columns correspond to scenarios for different management options, and the rows to different aspects of the problem. Stakeholders would be asked to colour in the cells of the table, using the coding shown, to indicate how they see each combination of sub-issue and scenario. Then the tables are collected and displayed in some way, either with or without names. The 'votes' could also be combined by software to display blended colours. The purpose is not to reach an immediate decision, but to display and discuss these preferences. The tool is most useful when it enables people to see what their neighbours think, and as a result to consider changing their own minds and be able to see these changes over time, for example, in a series of meetings.

A more complex, software-based, deliberation support tool is shown in Figure 5.4. The identified coastal problems are the 'Issues 1, 2, 3' and the possible model-generated futures are the 'Scenarios 1, 2, 3'. The new, third dimension, which makes this matrix a cube, are the different types or groups of stakeholders, labelled here the 'Actors 1, 2, 3'. Stakeholders vote on each combination of issue and scenario, and then the software displays the aggregate judgements in the colours of the balls or bubbles, including mixed colours where there are several opinions.

Such information is difficult to read and understand when presented in three dimensions, and the complexity of the 'cube' shows us that deliberation processes will in most cases be too complicated to simulate by a model, as they will be unpredictable mixtures of cognitive processes, emotional intensity, social hierarchies and power relations. However, the software allows the cube to be inspected one face at

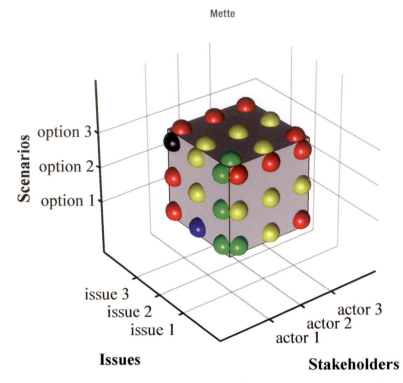

5.4 Elements of a Deliberation Matrix tool. The elements represent the three dimensions of such matrix: an axis for the scenarios of possible futures (issues), an axis for categories of stakeholders (here called 'actors'), and an axis for the Issues. From Mette *et al.* (2011), see also URL: http://kercoasts.kerbabel.net/.

a time, simplifying the bubble-clutter. It can then be turned so that the bubbles for actor-issue (which issue is more important to whom?), actor-scenario (which scenario seems better for whom?) and scenario-issue (which scenario seems best for which issue?) can be seen.

Thus the cube serves as a focus for discussion, informing people about the views of the stakeholder collective and facilitating the process of moving towards consensus, or, at least, seeing that consensus is unobtainable. Furthermore, setting out issues, scenarios and actors in this way helps to demonstrate the nature and dimensions of a problem, and may help the shaping of new scenarios. The opportunity for stakeholders to compare their own opinions with those of others, in a semi-anonymous way, not only makes it possible to discover the diversity of points of view (or their coherence), but also to reflect on and possibly change initial views.

Deliberation, informed in this way, is therefore social learning in the form of a dialogue on the comparative acceptability of different scenarios and their outcome (Mette *et al.*, 2011). In a real-life deliberation, with disagreements, the objective is to see what potential there is for agreement between the different interested parties about which possible future to go for in their coastal zone. Even when there is no perfect agreement, the majority party may come to appreciate that they must compensate the minority party for the disbenefits that might fall to them if the most popular scenario is put into effect.

In a real coastal policy-making situation, the use of such tools will in most cases be tied in with either a coastal area, or a maritime, planning process – or with policy decisions related to permissions for harvesting of a certain coastal resource – or for cultivations in a certain coastal ecosystem.

Making a SAF application

Just to remind you: a Systems Approach is about systems and models, but it is also about people in the real world. In an application of the Systems Approach, transdisciplinarity, deliberation and communication methods help to improve the available information and knowledge, to include communication and participation channels which cannot be put into a mathematical model, and as a result of this to nourish the science policy integration with preferences, opinions and interpretations; i.e. with qualitative information (see also Figures 1.6 and 5.2).

Let's get started

What happens at the start of things? When we began to think about explaining to our colleagues how an application of the Systems Approach Framework might start, we thought of a fairy story:

> Once upon a time – our tale might begin – there was a room full of scientists. Not stakeholders or policy-makers, but people who had been trained in a special way of knowing things. There was an ecologist looking down a microscope, a modeller typing at a computer keyboard, two social scientists discussing whether spending time on Facebook counted as building social capital, and an economist worrying about how all this was to be paid for. Suddenly, they heard a noise at the door, and in staggered a stakeholder, a black-feathered arrow labelled 'impact' protruding from his back. As they gathered round to pull it out, the scientists could see that a long thread linked the arrow to something quite distant, labelled 'human activity'. 'You know what this means?' said the one in charge, 'a dysfunction in the social-ecological system. Saddle up, men (and women), for we have an Issue to identify.'

In reality, the application might be initiated by stakeholders who seek better information to help them choose amongst management options already proposed, for example, by regional planners. Or the starting gun might be fired by local environment managers, who have themselves identified an environmental problem, or know that they soon have to implement a new law, and would like more information about the consequences of their planned actions. Or scientists themselves may start the process, through their own concern about an environmental problem. In many cases the kick-off will be a messy process, involving repeated meetings between the three groups of actors, during which the essence is slowly distilled from an initially confusing set of problems, perceived impacts, and potential solutions. And it may be the case that the researchers are called onto the scene only after the policy-making has taken place, in order to help with implementation. In that case, it is likely that none of the decision-makers has thought about whether or not civil society would agree to support the policy (Borner, 2011) – or perhaps they have thought too much and decided that participation was not wanted. And in the worst case, the catastrophe has already occurred and the social-ecological system is teetering on the brink of collapse.

Well – whatever initiates the process, the recommended sequence of activities remains the same. The SAF application should start with an analysis of the problem, or a problem framing, a task that can also be called systems-analysis or stakeholder-issue-mapping. The key here is to describe and illustrate a social-ecological problem, which we will call the Issue, and to identify the stakeholders involved as well as understanding their interest concerning that issue. To set this key knowledge in a broader context, it is wise to make a detailed list of human activities and associated stakeholder groups. Such a list is exemplified in Table 5.1.

It is also a good idea to make an institutional map which helps with under-standing how the governance relates to the human activities and the stakeholders. As learned in Chapter 2, institutional mapping is a 'procedure for identifying socio-economic relationships amongst institutions, organisations and groups' (McFadden et al., 2011). Figure 5.5 shows part of such a map, for the Issue of eutrophication in the Swedish Himmer fjord. We begin to see the complexity involved in the Issue, even while still far away from formulating mathematical functions to represent the ecological and socio-economical dimensions of the system.

Table 5.1 Stakeholder-issue mapping. Several ways to look at the stakeholder groups relating to the policy Issue of eutrophication in Himmerfjärden. Made using the CATWOE tool (Checkland, 1999).

	Agriculture (demand for less leakage)	Agriculture (conventional)	Agriculture (bio-dynamic)	Sewage Treatment Plant	Summer housing (nutrients from private sewers)
Human Activity (related to the Issue)					
Customers (who benefit or suffer from the transformation)	Benefiters: Recreationists, publics? Victims: Farmers? Taxpayers?	Benefiters: Landowners/ farmers & public Victims: Publics, recreation	Benefiters: Public, recreation Victims: Farmers?	Citizens in southern Stockholm	Beneficiaries: owners of summer houses; Victims: publics, tourism
Actors (who carry out the transformation)	Landowner/ farmers	Landowner/ farmers	Landowner/ farmers	SYVAB	Owners of summer houses
Transformation (which converts inputs to outputs)	Conventional land use plus leakage minimisation (through constructing wetlands?)	Supplying demand for agricultural goods	Supplying demand for biodynamic agricultural goods	Supplying requirement for STP for citizens in Stockholm	Supplying demand for tourism/ summer housing
Worldview (the bigger picture)	Public concern for the aquatic environment. (Farmers have willingness to cooperate with each other and with authorities)	Agricultural goods are necessary. Biodynamic goods are more/too expensive.	Biodynamic goods are better for the environment and human health and the additional cost is justifiable.	Sanitary/sewer facilities as standard.	Summer housing is common recreation in Sweden
Owners (who can start or stop the transformation)	Authorities, landowners/ farmers	Landowner/ farmers, CAP, government	Landowner/ farmers, CAP, government	SYVAB	Municipalities (through planning) and SEPA (through advisement and interpretations of the legislation)
Environmental Constraints (physical, financial, legal, etc)	Physical conditions (for e.g. wetland creation); national land use regulations and national/ EU agricultural policy	Physical condition and national land use regulations and CAP. Existing technology.	Physical condition and national land use regulations and the CAP. Existing technology.	The existing technology used in the STP, and legislation	Environmental regulation/ legislation, existing technology, information

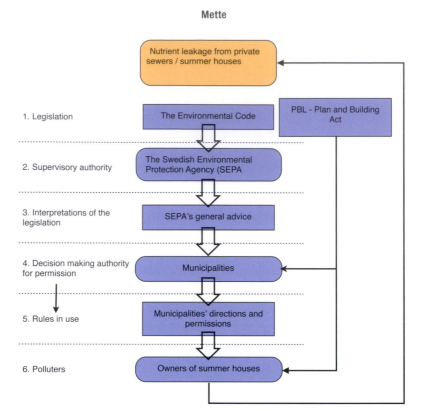

5.5 Part of an institutional map for the Issue of Eutrophication in the Swedish Himmer fjord. The SAF application team drew three institutional maps for the study site to represent the different scales and relationships. This part deals with private sewers, which discharge household waste directly into natural waters. These are only one of the sources of nutrients entering the fjord, and the map relates to only one of the categories of people who have a stake in the Issue.

Issue Identification

Equipped with these maps, and hence some knowledge of the stakeholders, issues, and governance, the engagement process needs a working group to proceed. This group ideally consists of stakeholders or their representatives, together with environmental managers or policy-makers. In some places, such a group might already exist, or can be recruited from an existing environmental forum. In other cases, this step of bringing the group together for the first time might be an arduous one. Imagine the very upset vessel owner in southern Denmark being invited to meet the environment manager who – from the vessel owner's point of view – destroyed the economic well-being of himself and his family by banning mussel harvesting during summer. Or those Venetian fishermen whose troublesome last meeting made it into the newspapers. There are better things to do than to go to such a meeting, they will think. But although it is possible to take account of stakeholder interests without the presence of stakeholders, to do so would risk misunderstanding these interests and would lose all the advantages that we have previously documented. So every effort should be made to inform them about the incentives for coming aboard, and being heard in, a SAF process.

Based on the first maps, **human activities**, impacts, management options, and indicators can be discussed with this **reference group** – so called because matters are referred to its members. The discussion should make use of participatory tools and – as was done in SPICOSA – a deliberation support tool, a DST. This stage is called Issue Identification because the goal is to reach a consensus on the matter to be addressed by the SAF application. The essence here is to focus in on, and define, the main environmental problem of concern, its impact and its cause; to agree on social-ecological indicators; and to identify possible management options. These management options will later become the basis for building scenarios to feed into a virtual machine; a virtual representation of the system, as you read in chapters 3 and 4. Be aware that 'reaching consensus' is an ideal and is to be understood in terms of the theory of deliberative rationality given earlier in this chapter. Furthermore, it does not mean that the problem of the impacts of human activities is *solved* by what follows – Issue Identification! Instead, think of this stage as beginning a local recon-figuration or adaptation of the social-ecosystem in the direction of maintaining or increasing sustainability, etc.

Table 5.2. shows a short summary of an example Issue, defined in the same Swedish fjord as before. The table shows 'water transparency' as an example of an environ-mental indicator. It is easy to measure, widely understood, clearly relevant to the issue of eutrophication, and allows the success of management options to be assessed. The socio-economic indicators will, perhaps, depend on stakeholder preferences. Is the aim to maximise local income, for instance? Or to maximise employment?

An example of a societal debate about the nature of 'the problem' can be found further south, where people are worrying about mussels in the Mar Piccolo, a shel-tered bay next to the city of Taranto in southern Italy. The city is an important commercial port as well as the main base of the Italian navy. Besides this, it is an industrial hotspot with the biggest iron and steel centre in Europe, yards for build-ing warships, and factories for petrochemicals, cement and food-processing. It is also one of the most polluted European cities. In 2006 Taranto suffered an economic crisis, accumulated a debt of 637 million euros, and the mayor went to jail. The crisis was complex and cross-sectorial, and the problems with mussel cultivation seem to reproduce this in miniature.

Table 5.2 Himmerfjärden, south of Stockholm. An example policy Issue.

Reference group:	About a dozen, including farmers, private citizens, elected representatives, officials from municipal authorities and the Environment Protection Agency
Human Activities:	Discharges from Sewage Treatment Plants, agriculture, and private sewers
Forcing:	Enrichment of the fjord with nutrients
Impact:	Degradation of water quality which can deter tourists
(Policy) Issue:	Eutrophication
Management options (scenarios):	(i) increased stripping of nitrogen from STW discharge; (ii) connection of private sewers to public STW plant; (iii) change in farming practices so that smaller amounts of nitrogen compounds enter the fjord
Social concerns:	Desire for clean water in fjord, distribution of costs amongst stakeholder groups.
Economic aspects:	Costs of sewage treatment, benefits of leisure visits
Provisional Indicators:	Water transparency, number of visitors during year

The apparent 'mussel problem' was a drastic reduction in the quantity, size and taste of harvested mussels, resulting in an impact on the local economy. It is possible that the cause of this reduction was decreased availability of food because of improvements in sewage treatment and discharge. Paradoxically, whilst sewage bacteria and toxins can contaminate shellfish, the organic matter in sewage can provide a source of food for mussels, and the nitrogen and phosphorus content can fertilise the growth of phytoplankton, also providing more food for the shellfish – an example of good intentions, and a good law (the Urban Waste Water Treatment Directive) leading to an unexpected outcome.

This example shows that the ecological part of the cause-and-effect chain may not be simple. But there is more to unpack. In addition to the economic difficulties of the mariculture, it also ran into social difficulties. The mussel farmers were getting older, but their concessions were not transferable and could not be passed onto their families. Because it was not easy for younger people to enter the legal activity, there has been an increase in illegal cultivation and illegal support services. There are about 10 small illegal docks of 10 to 30 boats with no efficient infrastructure – let alone services. The waste is being burnt and scraps are dumped into the water, contributing to pollution.

Mar Piccolo was SPICOSA study site 14, and the problems of mariculture were tentatively identified by the scientific team as appropriate for the resources they had available for a SAF application. Most of the stakeholders and managers who were invited to take part in the application were therefore involved in mussel culture. They included the mussel fishers themselves, delegates from industry, and representatives of the municipality, the Province of Taranto, an environment agency, the Harbour Offices and Authorities, as well as the [city's] Health Office. In a meeting of this reference group, and after a first round of deliberation, the Issue was defined as 'Sustainable use of the Mar Piccolo resources in order to include mussel culture' (Caroppo *et al.*, 2010).

The main questions that seemed to need an answer were:

a) What are the environmental conditions that control or benefit the mussel?

b) What would be the costs and benefits of enacting the measures needed for sustainable mussel growth?

c) What are the effects on human and ecological health resulting from the exposure to hazardous contaminants and organic wastes produced by industry and the illegal fishery?

So far, so good. But at this stage, it is worth pausing to reflect on both the process and the emerging components of the Issue, as these have been reported to us. Were these the result of reaching a consensus in the reference group? Were the scientists maybe steering, or trying to steer, the process? Wouldn't an environment agency or grass-roots organisation for the protection of the environment have views that differed from those of the representative of a chemical plant? And would the mussel farmers maybe not want to sit on one table with the representatives of the municipality? To all these questions, the likely answer is 'yes'. Nevertheless, it is apparent that the reference group, instead of splitting up in utter conflict, came to an agreed and rather precise outcome by focusing on intense communication, dialogue and deliberation.

So this is how our SAF version of an integrated assessment begins: by mapping stakeholders and institutions and by defining a policy issue and management options – ideally in a deliberative process. Even at this early stage, the scientific team begins to have a good basis for building a conceptual model, as described in Chapter 4. That chapter was pondering the arrow from the clam box to the clam fishery box in the model and whether to use energy or weight as measure for mussels. It tried to 'mathematicise' the work that is done in the issue identification step. But how is it possible to put the clam or mussel fishers' opinions and stakes, as well as money, the mussels, and micro-algal cells, into equations? How to make the equations reflect reality?

The Issue Identification step and the System Design step of a SAF application include answers to these questions. Before, or even while, building such conceptual model, the policy issue as well as the associated policy or management options, indicators, descriptions and criteria will need to be described and specified as well as possible. We don't start by writing 'X = *start value of mussel fisher Salvatore*', but instead begin by constructing a narrative about Salvatore and his co-workers and their families, and their part in the social-ecological system. This narrative will later provide the basis for the visualisations used in building the conceptual model – and its arrows. Its arrows are important, because they show relationships and forces.

To complete the specification of an Issue, we need to know several things:

❏ What dysfunction(s) in the natural system is implied by the chosen policy issue?

❏ Which cause-and-effect chains exist, arising from human activities, such as those of the mussel fisher?

❏ Which ecological indicators should be used in comparing the outcomes of the possible management options?

❏ Which are the economic activities and potential economic effects that are directly impacted? And which are the main economic drivers of change within the coastal zone system?

❏ Which are the main ecosystem goods and services relevant to the issue?

Table 5.3 Mar Piccolo, Taranto: List of the main Human Activities, Dysfunctions and Impacts.

Human Activities	Ecological dysfunction	Key variables linking forcing to impact	Impacts on ecosystems goods and services
Mussel culture	Reduction of the mussel productivity and health	Mussel recruitment, employment, local jobs	Reduction of the local market, loss of jobs
Urbanisation	Eutrophication effects	Anoxia, benthic habitat, diversity loss, toxic and harmful algal blooms	Change of trophic structure
Heavy industry	Biochemical pollution	Heavy metals, PAHs PCBs	Contaminated mussels, stress on organisms
Agriculture	Eutrophication and toxic substances	Fertilizer use, field drainage, crops, pesticide use, surface water transport	Change of trophic structure, ground-water contamination, ammonia emissions
Navy docks, large ship traffic	Physical habitat destruction	Shoreline development, resuspension from large ship traffic	Reduction of mussels recruitment, perceived environ-mental quality
Transport	Diversity loss and invasive species	Bacteria, phytoplankton and zooplankton, phytobenthos and zoobenthos, nekton	Perceived environmental quality, water transparency

The first of these is exemplified in Table 5.3, which displays a number of human activities, dysfunctions and impacts relevant to the case of the Mar Picolo. It might prove unnecessary to take all of these into account, but suggests the complexity of even this part of the social-ecological system, and also, exemplifies a preliminary conceptualization that should be further discussed with stakeholders, to benefit from their own knowledge of the system.

System Design, Formulation and Appraisal

We are now ready to flip back to chapter 4 and the activities concerned with the design of the conceptual model and the simulation model. The ideal starting point to the simulation model is a multi-layered, well-scaled, and well-visualised conceptual model. The layering deals with system heterogeneity and hierarchy, going from less detail to more detail. In most of the cases investigated during SPICOSA, the scale was local, dealing with an area the dimensions of a classical Greek city-state, placing larger scale entities and processes outside the system boundary, as part of the boundary conditions. And the visualisation allows checking with the reference group whether or not this representation of a system is coherent to their system – the real world in which they live their lives, whether as fisher, hotel owner, or environment manager. Such a conceptual model also supports the awareness of complexity and encourages the early detection of uncertainties and errors.

Gathering the information to make the conceptual model is part of what we have called the System Design step in a SAF application, the step that follows on from Issue Identification (Figure 1.5). We sometimes think of the SAF as a house; on the ground floor are the public rooms, where stakeholders come to visit and deliberation occurs. Moving through System Design to the third step, System Formulation, seems like going down into the basement, where the modellers live like mushrooms by the pale glow of their computer screens. This is to say that we are moving from steps in which the main challenges are the social ones, of engaging and working with stakeholders, to the technical ones, of constructing the virtual machine. But it is important not to separate these challenges too much. Part of System Design is a task called **Problem Scaling**, which requires the modellers to return to daylight and talk over their ideas for the virtual machine with their scientific colleagues and with the members of the reference group. Is the design too simple – perhaps lacking important components or links – or too complex – perhaps so detailed that it will be costly and time-consuming to make into a simulation model? A SAF application needs to be an iterative process, with recurring exchanges of ideas between science and society.

The next step, the one that will take place mainly in the basement, is System Formulation. It is so called because those things, that are fed into the model to explore the outcomes of scenarios, will need to be 'mathematicised' – put into formulae; or equations, or algorithms, or computer code – in any case, rendering the qualitative picture of the conceptual model into a set of strict rules for manipulating numbers. Once this is done and the multidisciplinary – or transdisciplinary – modelling team has built economical, ecological and (ideally) social modelling blocks, these can be connected – not with arrows any more but with little computerised connectors, switches, and of course filled with data, or linked to data-bases: data that are needed for the simulation of a chosen management option.

After this, it is time for the Appraisal Step. Looked up in a thesaurus, 'appraisal' can be substituted by 'assessment' or 'evaluation'. This has to be done indeed, because if we press the 'play' button of STELLA®, ExtendSim®, or other modelling software, we will see graphs and curves that show us what a future might look like. We now have to move carefully back to the 'real world', i.e. assess the simulation results and interpret them.

The first part of assessment is to investigate the reliability of the model results. As described in chapter 4, this typically means comparing the results of the baseline, or no-change, or hindcast, simulation with observations at the study site. In addition to this, an uncertainty analysis is desirable. We found, during SPICOSA, that this was likely to be requested by reference groups who have had a hand in System Design (Sastre *et al.*, 2010 and Franzén *et al.*, 2010). An *uncertainty analysis* is defined as follows:

> A quantification of uncertainty in model results due to incomplete knowledge of model parameters, input data, boundary conditions and conceptual model. In an uncertainty analysis the combined effects of these uncertainties are taken into account. (Gilbert *et al.*, 2011)

Interpretation, or **interpretative analysis**, is a technical and objective process, and refers to the translation or unpacking of the computer results in relation to the scenarios originally agreed with the reference group. It means to bring back together different graphs and curves, for example the simulations or derived values of mussel growth, economic growth, the unemployment rate, or the tourist's perception of water quality, to illustrate and explain the impacts of a certain scenario – impacts on the natural system, economic impacts, and the impacts on the lifeworld of *Homo sapiens littoralis*.

As before, this analysis may need to be an iterative process, with graphs tried out on stakeholders and then redesigned for greater clarity. But mostly it's a basement process, generating the 'ready-to-go' interpretation of the results for use in the next and final step.

The Output Step

Climate Change! The need to adapt to climate change intensifies the transition processes so that citizens can devote even less time to perception, interpretation, and evaluation of policy decisions. In a natural park close to the German Oder Estuary, the Marshland Protection Programme of the Federal State of Mecklenburg–Vorpommern for the re-naturation of marshlands is implemented for the protection of biodiversity and as a climate change mitigation measure. In 2010 environmentalists and tourists went on a walking tour to the now flooded polders. Reaching the polder landscapes, they were welcomed by farmers throwing tomatoes and pouring sewage. Similar incidents happen throughout many regions and show that if stakeholder groups do not accept the value and importance of such measures, they are able to obstruct sustainability policies. Taking a more optimistic view, stakeholders can be seen as key actors for building acceptance in civil society for the implementation of policies or management plans. For this, we argue, scientific knowledge, such as the existing knowledge about climate change, needs to be transferred to stakeholders and to the citizens in more graspable ways: using methods that connect the necessity of re-naturing marshlands, or other sustainability measures, to their own lifeworlds.

Let's assume that the model results that we have exemplified from Venice, Barcelona, Taranto, Himmerfjärden, and other sites in the coastal zone, are now interpreted and appraised – ready for delivery back to the real world! In chapter 3, we often referred to systems as 'black boxes'. Thinking in that way, we have described how stakeholders provided the input to the system and checked what the scientists were putting in the box. Now there is some output. It has to be translated for each of the several groups that might make up the audience for the first part of the Output Step.

It belongs to the concept of transdisciplinarity to make information from research accessible to the public, and that is in part what this step is about. We want the stakeholders, etc. to hear and understand the information that science has provided. Stakeholders and other reference group members who have been engaged during the whole process of a SAF application will be prepared and receptive for such understanding. They will already be familiar with scenarios and the idea of running simulations using models of virtual systems. It will be harder work for the scientists and the audience in cases where there have been non-engaged stakeholders – especially in the worst case, where an Output step meeting is the first time that stakeholders have had a chance to interact with managers and scientists. Nevertheless, the same principles apply in all cases.

But there is more to the Output Step than merely communicating results. The aim is communication as part of social learning, leading to deliberation and then, we hope, to improved decision-making by whoever is charged with making decisions. Thus, it is important to recall the communication skills that will be needed again in this new iteration of the engagement and deliberation process, as well as in explaining and translating scientific results.

A key part of System Output is that of communicating the modelling results as scenarios for possible futures, and deliberating on these. Working with scenarios sometimes requires a paradigm shift. Northern cultures tend to think prognostically, in terms of planning, and cause and effect chains – as if there is only one future available, and as if, given enough data and knowledge, we can see it running before us like tram tracks.

The use of systems scenarios and their underlying models opens the way to a new learning and thinking culture. Thinking in possibilities! When working with, and thinking in, 'what-if' scenarios, there are options to shape or influence the 'what' – potentialities that we earlier called ownership or empowerment. If this influence exists, then the future is decided upon, or at least partly shaped by, society. Ulrich Golücke (2001) says that 'scenarios are stories of the future that motivate people to do something'. Alternatives can show both the success that can be gained, and the price that has to be paid. Unlike utopias, scenarios should be practical and linked to the here-and-now, to existing social contexts, to trends and to developments of the region or community. They can be hooked to technology and innovations which are considered as possible/feasible for expert groups. The presentation of simulations that display several scenarios is therefore a powerful means of focusing stakeholder discussion and deliberation on the most realistic and effective measures available. The legitimacy of decided measures is likely to increase if these are supported by the simulation results presented.

What is presented in the Output Step should reflect all the earlier steps of the SAF application, including its debates. The scientific findings should be shown with as much transparency and scientific honesty as possible: i.e., with open arguments for the adopted scientific perspective within the model. Assumptions and methods should be clearly articulated to the forum or reference group, as should any conflicts between long-term sustainability concerns and shorter-term management or policy-making issues.

Visualisation

Again: Communication! Scenario presentations provide the main theme for the Output Step, but communication is still the key. The conceptual model and the mathematical model will only be of value for sustaining our coastal zones if they are understandable to those who have to live with the outcomes of a management decision that might be made. Of course, this refers only to the humans in the social-ecological system. We expect the puffin and the salmon, the mussel and the clam, or their analogues, to be integrated by now into the SAF's virtual machine. The communication of complexity and of the interpretation of the scientific information is the last crucial task of a SAF approach, leading to a final deliberation.

Back to the contested topic of climate change for an illustration. Whether one agrees with Al Gore or not, his movie (Guggenheim, 2006) presents an extremely complex global issue in such a way that lay people are able to understand and grasp it. When using graphics, he always finds an adequate scale and way to show them. It might seem almost ridiculous for him to appear on a lifting platform to illustrate different dimensions and scales. But he had success with it! It was an immense achievement to translate the data of the Intergovernmental Panel on Climate Change (IPCC) into public language. At first, the movie was criticised by experts for oversimplifying, but, together with the translation of the issue in the Stern Review on 'The Economics of Climate Change' (Stern 2006), it has initiated a public agenda for the discussion about climate change.

Gersmann (2010) and Borner (2011), see evidence that the issue of climate change has been brought to a wider public because of these two translations of scientific knowledge. This is not to suggest that it is necessary to prepare for scientific communication by running for President, or, like Nicholas Stern, taking charge of the World Bank. Nor is it necessary to make a hit movie or release a 700 page report. But it does suggest that it is important to think carefully about the needs, background and skills of the audience.

The Pertuis Charentais are waters on the west coast of France, protected by the Ile de Ré and the Ile d'Oléron. 'Pertuis' is usually translated as 'strait', and there are three of these, the central and largest Pertuis d'Antioche running from the Atlantic south-eastwards towards the mouth of the Charente river and the Basin de Marennes Oléron. The latter is famous for its cultivated oysters. At SPICOSA study site 10, the stakeholder group together with the environment managers and scientists have focused on the issue of quantitative management of the freshwater in the Charente river catchment. Local farmers need river water for crop irrigation; river water is also needed for households and industries; recreational fisheries depend on good flow; and the success of oysters in the offshore basin depends on the marine circulation induced by river discharge. Thus, low river levels during summer impact on a number of services provided by the aquatic ecosystem, and can lead to conflicts between different human use-groups.

Figure 5.6 is a top-level conceptual model showing some of the relevant links for the Charente river catchment and its coastal zone. It includes not only the ecological aspects but also institutional frames such as EU water and agricultural policies and national laws establishing hierarchies of water uses and setting volumes that can be used for irrigation. Incidentally, it is in French. We left it that way to give you (if you don't read French) a hint of how a scientific diagram looks to a non-scientist. But there is some help in the legend. The figure was made with software that allows conceptual objects to be opened up, revealing additional detail or lower levels in the system hierarchy. But even the top level gives an idea of the complexity likely in a virtual representation of the system.

Figure 5.7 is an example of smart visualisation of a simulation model in the Output Step. It shows the water flow level, calculated by the modelling software, interpreted in relation to irrigation rules as coloured lights. If the light turns red, the farmer won't be able to irrigate. Traffic light green: irrigation possible. The smaller part summarises outcomes in one locality under two different scenarios for the irrigation rules; and these scenarios could relate to the provision of river discharge into the Marennes-Oléron basin, which is an important variable for the oysters.

Visualisation and translation into the language of the audience are two core parts of the Output Step. The third one is to work with scenarios: to show the costs and benefits of each management option for the stakeholders and the environment, explaining uncertainties and assumptions. These possible futures can then be compared and used to initiate another round of deliberation. This time the aim is not to identify an issue but to wrap up the SAF-loop, if possible by reaching a consensus about what should be done, or by at least informing the decision-makers about stakeholder opinions, now properly informed by the scientific exploration of the scenarios. All this should help build a more informed and a more accepted policy-making process in the coastal zones.

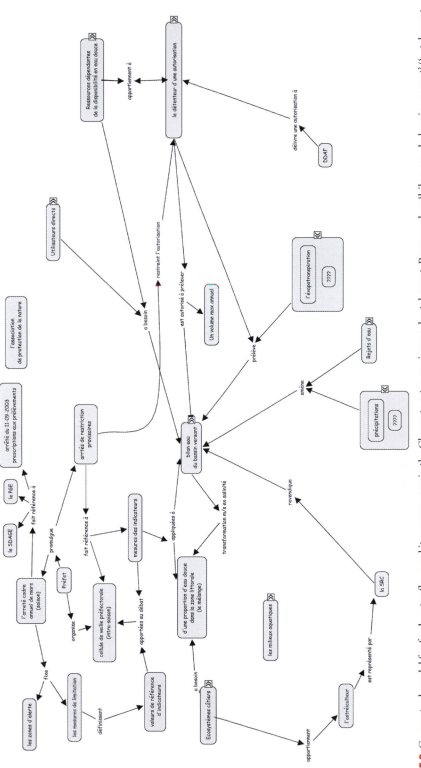

5.6 Conceptual model for freshwater flow and its governance in the Charente estuary, river and catchment. Boxes such as 'bilan eau du bassin versant' ('catchment water balance') contain unshown details of the system. 'Un arrêté' is a public decree or regulation; 'l'arrêté cadre annuel' is issued yearly to provide a framework for restricting use of water.

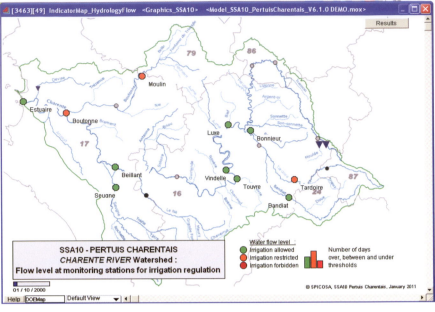

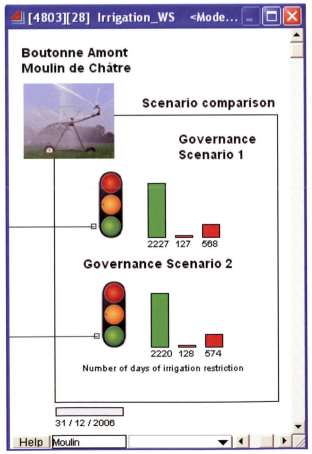

5.7 Screens from an ExtendSim® model. Animated mapping of Charente river water flow level at monitoring stations for irrigation regulation. The traffic light visualisation shows to what times irrigation is allowed and where.

Conclusion

> Application of the SAF allowed the stakeholders to influence the research, hence increased the likelihood that the decided measures will be seen as reasonable and effective, and also established a common learning platform among stakeholders, including managers, and the researchers (Franzén *et al.*, 2010).

This quote was taken from the final report of an application of the SAF at the SPICOSA study site in Himmerfjärden. It documents the importance of stakeholder engagement and the social learning process in which all the participants, including scientists, took part. Equipping the participants for further learning can be an important outcome of a SAF application, because it is desirable to communicate clearly that policy or operational solutions reached after one application of systems modelling and scenario deliberation should not be regarded as the final answer or last word. There is no universal remedy for all environmental ills in the coastal zone. Instead, by taking a systems approach to policy-making, we can explore, understand, and harness the small but feasible changes in human activity that together can make a difference to the ecological, economic and social sustainability of coastal zone social-ecological systems. The added value for policy-makers is that, if scientists can function as an early warning system for social-ecological problems, problems are pointed out to them earlier and faster.

Were they equipped with writing or voting skills, or at least some abilities to communicate with our social systems concerning the impacts of human activities and the ecosystem's dynamics, the travelling salmon and the little puffin in chapter 2 would most probably favour such an integrated approach and strongly vote for a permanent and effective dialogue between scientists, policy-makers and stakeholders.

EndNotes

[Introduction – the need for communication] The term *soft science* distinguishes the rather qualitative data and methods of the social sciences from more quantitative data and experimental methods of the *hard sciences*. The challenge for the SAF is to integrate the 'hard' quantitative data of the natural system with the sometimes 'soft' qualitative aspects of the social system. The latter cannot always be simulated, but can by no means be ignored. *See also* 'hard and soft systems' in chapter 3. As used here, the term *ownership* means, especially, that people or organisations identify with and take on or share responsibility for a process such as sustainable development (German Ministry for Economic Cooperation and Development, 2011) or a

SAF application. It also implies agency, a capacity for (relevant and willed) action. Compare with the definition that is part of *CATWOE* (*see* below).

[What do you need for a SAF application?] Scientists can exhibit bias as scientists, because they need to secure money or resources to carry out their investigations or advance their careers.

[Participation tools] The word *code* originates from semiotics. It is used for conventions in communication, which are rules, constraints or restrictions – spoken or unspoken – that influence the meanings and interpretations of signs used in a text, a presentation, a photo, etc. The 'production and interpretation of texts depends upon the existence of codes or conventions for communication' (Jakobson, 1971). The 'reception of any … message is dependent on prior knowledge of possibilities; we can only recognise what we know' (Chandler, 2001). Codes are 'interpretive frameworks' (Chandler, 2001) and not only used by the sender of a message but – most importantly in our case – by the recipient. See also Eco's *Theory of Semiotics* (1976).

[Deliberation and supporting tools] Guidance in organising and running deliberative meetings is given by SustainAbility Ltd (1996). **Stakeholder forums** are, commonly, bodies of stakeholders who meet regularly to exchange views. Often the members are representatives of public and private bodies who see themselves as the interested parties; some of these bodies may be made up of stakeholders according to our usage of the term; others are companies who have a stake as legal persons in some jurisdictions, and others belong to 'governance', and their stake is that the organisations have a duty to regulate environmental quality. The DST we describe was developed by the REEDS research centre, University of Versailles Saint-Quentin-en-Yvelines, France, and is available at URL: kercoasts.kerbabel.net (accessed 31 January, 2011). *See* Douguet *et al.* (2011).

[Let's get started] **CATWOE** (Checkland, 1999), used in Table 5.1, is a tool for analysing a system problem or process into Customers (who are affected by it), Actors (who cause or prevent it), the Transformation process (which causes it by converting system inputs to outputs), a Worldview (the perception of T that makes it meaningful to C, A and O), Owners (who can start or stop the process) and Environment (the relevant out-of-system elements). Use of CATWOE can lead to more than one analysis of a real system. *See also* URL: creatingminds.org/tools/catwoe.htm (visited 1 February 2011).

[System Design, Formulation and Appraisal] Multidisciplinarity 'juxtaposes disciplines. [It] fosters wider knowledge, information, and methods. Disciplines remain separate, disciplinary elements retain their original identity, and the existing structure of knowledge is not questioned.' (Frodeman *et al.*, 2010). In this book, we refer to multidisciplinary teams because groups who apply the SAF must consist of scientists from several disciplines. Transdisciplinarity implies a merging of the disciplines, ideally generating a synthesis with a new outlook.

[Output Step] The story about the protesting farmers was recounted to the author by someone who had been involved. Not all farmers were upset; some of them received subsidies for re-naturation.

CHAPTER 6

Conclusions

Paul Tett, Anne Mette, Audun Sandberg, Marta Estrada, Maurizio Ribeira d'Alcalà, Tom Sawyer Hopkins and Denis Bailly

The problem

Much of humanity, the part we have whimsically characterised as *Homo sapiens littoralis*, lives in the coastal zone, the 'long narrow boundary between land and ocean that is a dynamic area of natural change and of increasing human use'. The interfaces between land, sea and air, the geological variation, and active geomorphology, make this a complex borderland, home to a variety of **biomes**, washed over by fluxes of materials and energy, and often highly productive of food and other natural resources. Humans have long found the zone an attractive place for settlement. But the needs of cities far outstrip the capacities of the local ecosystems to provide services; bringing in most of their food and raw materials from far away, they can easily overwhelm the **assimilative capacities** of local ecosystems with their discharges. And even villages and towns over-harvest the surrounding natural systems, burden them with wastes, and disturb their existence.

This is indeed a problem for the environment: for the natural ecosystems that are exploited, polluted, trampled on and overrun. But it is also a problem for human society, because of damage to the ecosystem services that the social systems depend on. Thus the problem is one of dysfunctions in social-ecological systems, dysfunctions that almost seem to diagnose a design fault in human society. If so, the fault is getting worse, because human resource consumption is increasing whilst resource assets are decreasing. Superimposed on this deteriorating situation is the hazard of rising sea levels, which puts much of the coastal zone, and many of the world's largest cities, at risk.

In short, stress on coastal systems is the acute problem for *Homo sapiens globalis*, even if many of the problems – ocean acidification, greenhouse effects, removal of higher trophic levels, etc. – impact the Earth system on larger scales. In this book we have mostly zoomed in on problems and solutions at the local scale, although sometimes pulling back to look at the planet as a whole, or to consider general aspects of social and ecological theory that we claim apply on all scales.

A solution

The argument of this book has been that a solution is built into what we have called *Homo sapiens societatis*. The species label, *sapiens*, given by Linneaus himself during his great, systematic endeavour to name all creatures, means 'wise'; our own modifier, *societatis*, means 'of society'. Our point is that humans both make, and are shaped by, social systems. Out of this wisdom and sociality emerges people's ability to engage in collective rationality and to create rules and institutions – including systems both of steering and ruling and of understanding – for environmental management. A pessimist, viewing the world, might conclude that it is dominated by conflicts of interest and that rational collective actions for ecological sustainability are uncommon. Nevertheless, this book is about a way of exploring such conflicts that can lead to a positive-sum outcome.

The book is also the result of a social learning process – about matters ecological as well as social and economic – by a group of scientists who took part in a European research project to test a systems approach to coastal zones. The book is not about the project, although we have used it as our main source of documented examples. It is about the methodology involved in a Systems Approach Framework, and the understanding required to apply such a SAF to coastal zone dysfunctions – an understanding that we claim is also useful for thinking about solutions to larger-scale problems.

A Systems Approach involves seeing coastal zones as social-ecological systems. Its methodology has two main components. The first component is a view of the interface between *science* and *policy* – that is, between scientists and policy-makers or environmental managers. This component seeks the active engagement of stakeholders in the identification of the dysfunction and in deliberation over possible solutions. We have argued that this is both fair (to stakeholders) and effective (in sharing with them the responsibility for maintaining sustainability).

The second component involves a synthesis of General Systems Theory and Soft Systems Methodology in designing a virtual system that involves a cause-and-effect chain from a human activity to its impact on ecosystem services. We write 'chain', as if the links were all linear, but in fact most coastal zone systems are more complicated than this. They include spatially- and scale- defined subsystems and feedback loops (chapter 3), giving them complex behaviours that must be represented – at least approximately – within the virtual system.

By way of one or more conceptual models, the virtual system forms the basis for simulation models (chapter 4) that allow the exploration and demonstration of scenarios, or possible futures. Some of these futures will include remediation of the dysfunction and maintenance or improvement of the sustainability of the ecological and social systems. The *efficiency* and *equitability* of model-derived solutions can be assessed by socio-economic appraisal using methods such as cost–benefit and multicriteria analyses, with full account taken of non-market costs and benefit. The latter is crucial, because otherwise decision-making will be badly informed about long-term sustainability. Returning to the stakeholders at this stage allows qualitative information – for example, about social values – to be brought into play (chapter 5).

René Descartes, in the *Discourse on Method* of 1637, made his second rule 'to divide each of the difficulties I examined into as many parts as possible and as may be required in order to resolve them better'. This is the analytical method, which has proven so successful for the development of scientific knowledge. Seeing things as a whole, but in detail, is more difficult than focusing on a part. Nevertheless, we've argued in this book that the complex network of interconnections found in social-ecological systems (whether in the coastal zone or elsewhere), full of feedback loops, hierarchies and emergent properties, cannot be fully understood by analysis. Only by conceptualising coastal zones as systems can scientists, officials and citizens properly understand their zones' tardy, exaggerated, or counter-intuitive responses to human activities, or comprehend the risks of walking them too close to the edge of the cliff in Figure 2.2.

The need for a Systems Approach

A case for a Systems Approach has been made in the preceding section, but let us back-track a little. If the natural environment is seen as something distinct from the human world, and if society agrees, for the urgent reasons set out in the first section, that it should be defended, then why not protect it with laws against pollution and over-exploitation? As we said in chapter 1, many countries have implemented environmental protection legislation. For instance, the European Urban Waste Water Treatment Directive of 1991 regulates the discharge of sewage. This contains both organic wastes, which absorb oxygen during their decay, and nutrients, which can bring about 'an accelerated growth of algae and higher forms of plant life to produce an undesirable disturbance to the balance of organisms present in the water and to the quality of the water concerned' – i.e., that can cause eutrophication. Implementing such directives is a process that can take more than a decade, but there is now evidence of reduction in the discharges of nutrients in the North Sea on that timescale (Vermaat *et al.*, 2008). However, urban waste water is only one source of nutrients; another Directive aimed to regulate the leaking of nitrates from farmland, and the OSPAR Strategy to Combat Eutrophication has also played a part in this reduction. We have already mentioned (in chapter 2) the potential conflict between the aim to reduce nutrient loading of the Danish Lim fjord and the beneficial effects of the additional food production for wild birds, which are protected by other European Directives, and for a new mussel fishery. And even before the UWWT and Nitrates Directives were issued, it was pointed out that organic matter in estuaries acted to remove nitrate-nitrogen through the process of denitrification; and thus cleaning estuaries of organic waste might increase their nitrate discharge (Billen *et al.*, 1985).

It was an awareness of these sorts of complex interactions that led to proposals for Integrated Coastal Zone Management, ICZM. In 2000 the European Commission recognised both the need for such an approach and the existing body of experience in applying integrated management. The Commission's 'Communication ... to the Council and European Parliament on Integrated Coastal Zone Management' summarised the problems of the coastal zone much as we have done in this book:

> [European] coastal zones are facing serious problems of habitat destruction, water contamination, coastal erosion and resource depletion. This depletion of the limited resources of the coastal zone (including the limited physical space) is leading to increasingly frequent conflict between uses, such as between aquaculture and tourism. Coastal zones also suffer from serious socio-economic and cultural problems, such as weakening of the social fabric, marginalisation, unemployment and destruction of property by erosion. ... [There are] many interrelated biological, physical and human problems presently facing these zones. Their cause can be traced to underlying problems related to a lack of knowledge, inappropriate and uncoordinated laws, a failure to involve stakeholders, and a lack of coordination between the relevant administrative bodies. There is no simple, legislative solution to these complex problems.

Therefore there was, and remains, a need for ICZM. And yet it has proven difficult to implement truly integrated strategies. Shipman and Stojanovic (2007) saw four kinds of difficulty:

> (1) the complexity of responsibilities at the coast continues to prevent agencies from taking a 'joined-up' approach; (2) a policy vacuum is constraining implementation from national to local scales; (3) informational obstacles are significant in preventing co-ordination between science and policy-makers, and between different sectors; (4) a democratic deficit is preventing implementation in the working practices of coastal stakeholders, with little opportunity in decision-making for public comment or local accountability, especially offshore.

Chapter 2 in this book has explored points (1) and (2). The Systems Approach Framework of this book is an attempt to deal with points (3) and (4). It addresses the Science–Policy interface with the conceptual model shown in Figure 1.6 and brings stakeholders into this interface by engaging them in identifying Issues and deliberating scenario-based solutions to dysfunctions. But, following Ostrom (2007) and others, we argue that the big remaining obstacle to good coastal zone management lies in the erroneous perception that the natural environment is distinct from the human world, and thus that we need to re-conceive human life as existing within social-ecological systems.

Does the Systems Approach work?

What is the evidence that it works, the approach recommended in this book? We have evidence from the studies carried out by the SPICOSA project, and documented in papers cited in this book, that the benefits are as much social as ecological. Stakeholder engagement is difficult but does appear to work. Not surprisingly, it goes better when there is an existing tradition of scientists working with governance and civil society – where there is existing social capital to build on, in the form of existing networks of trust and familiarity with each other's standpoints, stakes, and languages. Conceptual modelling has proven a useful communicative tool – allowing stakeholders and policy-makers to input their social and ecological knowledge into models – as well as a technical preliminary to simulation modelling.

Most of all, we find that taking part in a SAF application can lead to social learning. This is more than 'learning together'. It is the process and outcome of working together to analyse and conceptualise a problem, during which the understandings and practices of the participants change, whether they be policy-makers, stakeholders or scientists. And it is this change in the social part of the social-ecological system that we argue is the prerequisite for achieving sustainability in the ecological part of the system.

Such social learning, underpinned by the conceptual model of Figure 1.6, can be contrasted with a 'push'-model of scientific engagement, in which scientists pose and test hypotheses concerning an issue, and then teach (sometimes we say, politely, 'transfer') our findings to end-users. Some of us started with this push-model, perhaps seeing stakeholder engagement only as a way of getting people into a room where we could lecture them, whereas the experiences of writing guidance for SAF application teams, and of engaging with stakeholders and governance, has taught us much about the role of science in society and changed our scientific practices.

One important conclusion is of the need for transdisciplinarity in the scientific team. The very contents of this book demonstrate why a SAF application must involve expertise from a range of scientific disciplines. Interdisciplinarity brings together such experts and equips them to talk with each other. But ICZM, and a SAF, require a dialogue between science and society and science and policy. Science holds a special type of knowledge, but relevant expertise also exists outside university buildings and science laboratories. Look at G.D. Brewer's (1999) famous sentence: '*The world has problems, universities have departments*'. Many times quoted, this statement led to the assumption that science, organised in disciplines, does not give satisfactory answers to questions of social relevance. Transdisciplinary research integrates non-scientific expertise into all stages of the research process, which can help bind stakeholders and lay people into the research and its outcome, and lead to an increased legitimacy for the outcomes. Transdisciplinarity, we have learnt, is key to connecting scientific knowledge (based on hypothesis-refutation epistemology) and practical (or experience-based) knowledge (Bergmann & Schramm, 2008).

On the ecological side, we cannot point to any clear-cut examples of increased sustainability resulting from local application of a systems framework during the SPICOSA project. This is partly because the management of sustainability generally needs to persist beyond the duration of a typical research project or a conventional investment project. And although hard scientists – and their funders – would only be convinced by an experiment in which sites where the SAF was applied showed significant improvements over control sites, the softer view is that of the argument we made above: that long-term improvements in sustainability must begin in the social part of the social-ecological system. Therefore, we would want to judge the success or failure of the Systems Approach more broadly, and over a decade or more, by comparing sustainability in coastal regions where it is used with that in regions where it is not.

We are not claiming that a Systems Approach Framework is the panacea for solving all the emergencies in coastal systems. Nevertheless, we claim, and have tried to show, that using such a Framework has three important outcomes even when the dysfunction is not immediately remedied. The first is that, by focusing society's view on the social-ecological system, the SAF highlights the key elements that perturb its authentic dynamic. The second is that this broader view can lead to a reconciliation of competing forces at a level that goes beyond that which would typically be sought in the normal course of human events. The third is that even if a solution is not reached, an attempt of reconcilement reveals what we have still to learn about the system to manage it sustainably.

Science and Policy

We have mentioned the distinction between *Lifeworld* and *System* according to Jürgen Habermas. To simplify, the Lifeworld refers to the informal domain of social life, the place where communication amongst individuals reflects and shapes feelings, understandings and actions; the System is 'the sedimented structures and established patterns of instrumental action' based (in modern times) on money and power (Finlayson, 2005). Habermas sees the tension at the seam between these system worlds and the lifeworld of individuals. He values the lifeworld more than the system, on the grounds that it is often the former that drives change in society.

For us, using the somewhat different meaning of *system* explored in this book, both Lifeworld and System are subsystems of larger social-ecological systems. Money and Power are subsystems or institutions in Popper's world 3 (Popper, 1978), and they, as much as the lifeworld itself, depend ultimately on the services that ecosystems provide to satisfy the physical and mental needs of humans. Thus, as we have argued, human societies must be able to steer in the direction of ecological sustainability, and for this, large societies, comprising millions or billions of people, need effective institutions of environmental governance. And thus, the challenge for students of the relationship between science and policy is how to scale up from human-scale interactions to those necessary at the collective and constitutional levels of governance as implemented by democratic states and groups of democratic states such as the European Union.

We don't mean to argue that the methods in this book will work only in western-style democracies. Engaging stakeholders in human-environment management does not require that all citizens get to vote, and western-style democracies are not without problems. It is often difficult for politicians elected every two, four or five years to take actions for long-term sustainability if they are unpopular in the short term with their electorates. This is the matter of *timescales* that we mentioned earlier, and thus it is desirable to analyse system responses over several timescales, in order to show society that, sometimes, short-term benefits for a few can become long-term losses for the many.

When planning a science–policy integration, it is important to think back to chapter 2 and to be aware that any science–policy consultation takes place within certain institutional and governance frameworks, in which certain fundamental socio-ecological relations, such as cultural traditions or property rights, are very difficult to change and must be acknowledged, whereas laws and regulations can be changed after due political process, and constraints such as permissions, plans, or investment funding, can be changed more easily (Luhmann, 1985; Mette *et al.*, 2011).

Then, before satisfying our desire to bring science, policy and civil society together to deal with the fate of travelling salmon, sensitive puffins, and *Homo sapiens littoralis,* in more or less vulnerable socio-ecosystems, we have to remind ourselves of the roles we supply for these institutions. It is generally the role of policy-makers and environment managers to pick up and express the intentions of society. Science is the supplier of epistemologically well-founded knowledge and of simulations. It should not tell policy where to go and what to do. The methods in this book are intended to support science in illustrating consequences, impacts, scenarios – transferring its knowledge – whilst not undertaking the task of conducting political discourses. When the scientific work is done, and at which point it will be initiated – the impulse can either come from the policy-maker, the stakeholder, or the scientists themselves.

Charles Darwin told us that 'it is not the strongest of the species that survive nor the most intelligent but the most responsive to change'. Recent times have been characterised by rapid change. Since the nineteenth century, people have become used to economical and social evolution and, in some cases, revolution. But now there is a new and unfamiliar component, in massive ecological change. How is society to respond? In many parts of the world, the State increasingly stands back and takes a more enabling role, providing a framework within which civil society can organise itself, whilst aiming to strengthen the adaptive capacities of industries and government agencies. It may be that this frame will support stronger regional self-reliant processes – ultimately that of **autarky**. But autarky does not necessarily have to mean the little Gallic village of Asterix and Obelix against everybody else. Within Europe, the scale of such self-reliant regions could be imagined as being that of the Baltic area or the catchment and estuary of the Spanish-Portuguese Guadiana river. The former already has a strong transnational management institution in the Helsinki Commission, the governing body of the Convention on the Protection of the Marine Environment of the Baltic Sea Area. The European Water Framework Directive already mandates the joint management of river systems, such as the Guadiana, when they are shared by several nations. Although the EU lacks a powerful Coastal Management Directive, the new Marine Strategy Framework Directive moves in this direction, by adjoining adaptive management of member states' territorial seas to that already in place for rivers, estuaries and inshore waters.

The rest of the world

It is not only in Europe that coastal degradation and coastal vulnerability is increasing. In both North and South America, in Asia and in Australia/Oceania we find that the growth of coastal cities, the cutting of mangrove protective forest, the destruction of coral reefs and the depletion of local fish stocks is threatening vital coastal ecosystems and traditional coastal cultures. It is well known that certain forms of modernisation, e.g in the form of mass beach-tourism development, are increasing the vulnerability of coastal systems. On the other hand, there are large areas of the African coasts that are still robust and where the 'wild coastal ecosystems' are intact. In Southern Asia, the Brahmaputra and the Mekong Delta are not polderised like the major river deltas in the Netherlands, thus maintaining the adaptive capacity of the local populations to these extremely dynamic ecosystems. The appreciation that the construction of dykes, storm protections, flood harnesses, etc. can increase rather than diminish this vulnerability is increasingly entering the strategies of the large development agencies of the world: the the World Bank's International Development Agency, and the various UN development agencies. Thus the task of 'Ecosystem Restoration for Sustainable Development' is on the agenda in many non-European countries, emphasising the need to protect coral reefs and storm beaches from developers, regenerate mangrove forests, and 'give coastal rivers room' (Nellemann and Corcoran, 2010). In areas where the 'constructed coastal environment' is not as massive as in Europe, such ecosystem restoration is in many ways easier to achieve. In this kind of work, the systems-based approach discussed in this book can be a valuable help in identifying the most efficient sequence of tasks in a restoration process.

Table 6.1 Regional and Global Institutions concerned with the coastal zone or SAF-like methods.

Name: funding	aims	URL
LOICZ: global, 'UN-sponsored'	'... to provide science that contributes towards understanding the Earth system in order to inform, educate and contribute to the sustainability of the world's coastal zone'	www.loicz.org
EBM: USA-Australia network	'Human activities on land and in the ocean are changing coastal and marine ecosystems and threatening their ability to provide important benefits to society ... . Ecosystem-Based Management ... address[es] these challenges. It considers the whole ecosystem, including humans and the environment, rather than managing one issue or resource in isolation.'	www.ebmtools.org
ESPA: UK linked to 3rd world	'The Ecosystem Services for Poverty Alleviation ... research programme aims to deliver ... research that will improve our understanding of the way ecosystems function, the services they provide and the relationship with the political economy and sustainable growth. ... its goal is to ensure that, in developing countries, ecosystems are being sustainably managed in a way that contributes to poverty reduction and inclusive and sustainable growth.'	www.esi.ac.uk
MA: UN and 'informal global governance'	'The Millennium Ecosystem Assessment assessed the consequences of ecosystem change for human well-being. [Its] findings provide a ... scientific appraisal of the condition and trends in the world's ecosystems and the services they provide, as well as the scientific basis for action to conserve and use them sustainably.'	www.maweb.org

Table 6.1 lists some of the global initiatives pertaining to coastal challenges. The ICSU- and UN-supported LOICZ programme has traditionally been preoccupied with the degradations of the world's coastal areas, and LOICZ has worked with the IUCN for increased coastal and marine protection. Now, both these organisations have seen a need to work more proactively for sustainable development in areas where people live, and are promoting adaptive ecosystem-based management (EBM) in these areas as well as in more pristine regions. An important part of these new global development strategies is therefore increased efforts to empower coastal dwellers, who may be poor but often have valuable traditional ecological knowledge of how to reduce social-ecological vulnerability (cf. ESPA). The inclusion of hitherto marginalised and often illiterate groups of stakeholders poses special problems that have not been addressed in this book; but these problems are not insurmountable, and both the Issue identification and the deliberation aspects of a systems approach can easily be extended to a different social context.

Further resources

After reading this book you may be thinking about making your own SAF application or, at least, finding out more about some of the tools that have been mentioned here. Our description of the Framework has brought together a variety of methods as well as scientific and even philosophical concepts and theories, such as those of Popper, von Bertalanffy, Odum, Checkland, Holling, Luhmann, Habermas, Constanza, Ostrom, and Rolston – to name only a few. The reference list includes a number of textbooks for further reading, but we have found no single text that embraces all that we've dealt with here. We have tried to write that book, but our difficulty has been that of setting out all the information and ideas we think need to be displayed, and of organising it into a user-friendly, ready-to-go package. Our solution is to be found in another part of world 3, in the *wàn wéi wǎng*, the 'net with many links' or 'myriad dimensional net'.

At www.coastal-saf.eu (Figure 6.1) you will find an open source manual for the Systems Approach Framework. It is organised by the five steps of Figure 1.5, and is itself a systematic document. There is a practical top level that sets out the key work tasks of a SAF application, and deeper levels containing essential supporting materials, including introductions and in-depth information on economic tools and methods, a variety of examples, bibliographies and links to several databases and training modules. There you will also find some of the SPICOSA documents that we have cited in this book.

6.1 The Coastal-SAF website that provides further information on the topics in this book.

EndNotes

[The problem] The definition of the coastal zone is from LOICZ. It can also be argued that the the main dysfunction lies in the institutions of human society, which in their development lag behind the dynamism of the social-ecological systems.

[The solution] 'Fair to stakeholders'. We mean to suggest both objective fairness or even-handedness (although there would need to be agreement about how to measure this) and subjectively fair, as perceived by stakeholders and so leading them to see the relevant SADF application as legitimate. The word 'fair' is also meant to suggest that stakeholders have agency in, and some degree of ownership of, the process. *See* chapter 5.

[Does the Systems Approach work?] 'Social learning ... changed our scientific practice' Bremer (2010) has referred to this as 'post-normal science'. There is still no standardised, internationally-accepted definition for **transdisciplinary** research, but it is agreed that it goes 'beyond the academy' (Frodeman *et al.*, 2010). The characteristics of transdisciplinary projects are, on one hand, the crossing of disciplines (interdisciplinarity), and on the other hand, the interaction and cooperation amongst social, political and scientific processes of problem solving (Weingart, 2001).

[Science and policy] **Darwin**: the quote comes from a paraphrase of Darwin, probably by L.C.Megginson. The meaning is that species become extinct if they do not adapt to a changing environment. Neo-Darwinism understands 'species' as a gene-pool, in which the heritable information in genes is expressed in the phenotypes of individuals. Genes have different forms, or alleles. Natural selection acts on the frequency of these alleles by way of allowing the survival of phenotypes that are 'fitter', i.e. better suited to the current environment. Richard Dawkins (1989) argued for the role of the 'extended phenotype', exemplified by the dams made by beavers, in shaping fitness and being shaped by selection. In writing about **adaptability** in this section, we build on the argument of this book, that humanity's extended phenotype lies as much in world 3 as in world 1, and that selection can act on the 'inherited information' in world 3, favouring more resilient and adaptable social systems. Thus we are discussing 'adaptive capacity squared': the idea that part of *Homo sapiens'* adaptive capacity as a biological species is to form adaptive social systems, which can change without requiring further selection in the gene pool. But we go beyond Darwin and Dawkins, to argue that our extended phenotype must now include nature itself, because – so far as modern human civilisation is concerned – the key selective pressures act on social-ecological systems. **Enabling state:** *see* URL: www. oecd.org/dataoecd/7/34/35304720.pdf. **Self-reliant regions:** HELCOM, *see* URL: www.helcom.fi.

[The rest of the world] Sandberg (2010) documents agricultural risk-minimisation in a coastal delta and floodplain in Tanzania.

Glossary and Acronyms

For more and fuller definitions, see URL: www.coastal-saf.eu/glossary

adaptability ability of a system to reconfigure in response to change in boundary conditions (cf. resilience)

algorithm precise set of instructions for carrying out a sequence of calculations

assimilative capacity capacity of an *ecosystem* to absorb human wastes: part of *resilience*

autarky the property that results in a closed system being viable; usually understood as political and economic self-sufficiency

biome generic type of community of organisms (e.g. 'sea-grass meadow'), instanced in particular ecosystems

boundary conditions whatever lies outside the defined Virtual System and can influence the behaviour of that system

capital[s] include *natural* (e.g. fish stock), *social* (e.g. local networks of trust, institutions of governance), *financial* (money), and *durable* (e.g. infrastructure, manufacturing machinery)

carrying capacity ability of an *ecosystem* to support a stock (i.e. a *natural capital*)

CATWOE short for: Customer (who receives the results of T), Actor (who carries out T), Transformation (of inputs to outputs), World-View (the bigger picture), Owner (who can stop T), Environment (constraints on T): an acronym describing a Soft Systems Method for analysing problem situations and the human activity associated with them

code convention or rule in communication that influence the meanings and interpretations of writing, reading, speaking and hearing; what does not correspond to the reader's or hearer's code is filtered out

collective rationality process of making decisions together that produces sustainable social-ecological outcomes

communications space *see* Figure 1.6.

communicative rationality rationality that emerges after free communication and deliberation

conceptual model description of (a part of) reality in terms of words, equations, governing relationships or natural laws that encompasses the user's perception of the key processes

DEB dynamic energy budget

deliberation, deliberation matrix see *forum* and Figure 5.4.

DNA DeoxyriboNucleic Acid, the carrier of most heritable biological information

DPSIR Driver, Pressure, State, Impact, Response: *see* Luiten (1999)

ecological footprint 'measures humanity's demand on the biosphere in terms of the area of biologically productive land and sea required to provide the resources we use and to absorb our waste' (Hails *et al.*, 2008)

ecosystem 'area of nature that includes living organisms and nonliving substances interacting to produce an exchange of materials between the living and nonliving parts' (Odum, 1959)

ecosystem (goods &) services what *ecosystems* provide to humans

efficient (Pareto-efficient) state in which resources are so allocated that it is not possible to make anyone better off without making someone else worse off (economic objective)

emergent property *system* property that derives from component activities, but cannot be reduced to them

equitable perceived as fair by the members of a society (social objective)

EU European Union

eutrophic[ation] 'the enrichment of water by nutrients, especially compounds of nitrogen and/or phosphorus, causing an accelerated growth of algae and higher forms of plant life to produce an undesirable disturbance to the balance of organisms present in the water and to the quality of the water concerned' (UWWTD)

ExtendSim® (www.extendsim.com) simulation software made by Imagine That Inc, which uses an iconic interface

externality costs or benefits of the production or consumption of goods or services that are not figured into the prices charged

feedback [loop]: negative or positive closed loop structure that brings results from past action of a *system* back to control future action thanks to positive or negative effects that increase or decrease system rate processes

forum in classical (Roman) times, a market place also used for public meetings; in modern use, a place for exchange of information and for *deliberation*

goods and services *see* ecosystem

governance the steering and ruling of society and the ways in which citizens and groups articulate their interests, mediate their differences, and exercise their legal rights and obligations

GST General Systems Theory, which sees the world as containing *systems* with the properties given in Table 3.1

homeostasis self-regulation, the ability to maintain a near-constant *system* state despite external changes (*see* also: *feedback* and *resilience*)

human activity something that humans do that has consequences for the *social-ecological system*

IBM Individual Based Model

ICZM Integrated Coastal Zone Management

impact the consequences of a *human activity* for the social part of the *social-ecological system*, mediated through effects on *ecosystem services*; not the same as Impact in *DPSIR*

institution a set of rules and procedures, both formal and informal, that structure social interaction by constraining and enabling actors' behaviour.

institutional map diagram showing socio-economic relationships amongst *institutions*, organizations and groups

interpretive [or interpretative] analysis the unpacking, interpretation and translation of the results of computer simulation or of scientific observations of the real system

IPCC Intergovernmental Panel on Climate Change

Issue coastal zone problem or dysfunction that that needs attention from policy-makers or environment managers; during a *SAF* application the *impact* is identified and the Issue then comes to include the options and conflicts surrounding a policy or management action to mitigate this

Issue Identification the first step in an application of the *SAF*

Lifeworld (used by Habermas to describe) the informal domain of social life, the place where communication amongst individuals reflects and shapes feelings, understandings and action; cf. the *system*, 'the sedimented structures and established patterns of instrumental action' (Finlayson, 2005)

linear relationship a relationship in which the dependent (or output) variable changes in proportion to change in the independent (or input) variable. *See* Figure 3.3

LOICZ Land Ocean Interactions in the Coastal Zone, an international research project

MA or MEA Millennium Ecosystem Assessment, UN sponsored study of the consequence of ecosystem change for human well-being

market real or conceptual place where, and institution determining how, buyers and sellers interact to determine the prices of goods and services

model a simplified representation of the essential or dominant features of relationships amongst components of real systems, used to (i) increase and promote understanding of the real system, and (ii) simulate the behaviour of the real system under particular scenarios. *See* also: *conceptual model* and *simulation model*

MSFD (European) Marine Strategy Framework Directive: see Table 1.1

NOAA National Oceanic and Atmospheric Administration of the United States Department of Commerce

OECD Organization for Economic Co-operation and Development

ownership (of a process) being able to influence the process

parameter *model variable* that remains constant during a simulation

policy a deliberate plan of action to guide decisions and achieve rational outcome(s); mainly used of the legally binding plans of governance

policy-issue mapping tool for the identification of questions of a political nature that are perceived as coastal environmental problems and concern social-ecological dysfunctions

policy-maker individual, group or organisation with the power to propose, review and select policy options

policy–stakeholder mapping tool for the identification of actors (individual or organised) with a 'stake' (moral interest or legal right) in a policy issue

POM Particulate Organic Matter

problem scaling the scientific and management process of adjusting the complexity of a Virtual System so that it efficiently captures the relevant behaviour of a real coastal zone system relevant to an Issue

programme of measures management actions to achieve the objectives of the WFD in a water body or river basin (district) or of the MSFD in a marine region (or sub-region)

reference group small set of key stakeholders and public officials who take active part in a SAF application: to whom questions and results are referred

resilience capacity of a system to absorb perturbations, and so to persist without a qualitative change in its structure

SAF Systems Approach Framework, a methodology with 5 steps: *Issue Identification; System Design; System Formulation; System Appraisal*; and *System Output* (*see* Figure 1.5)

scenario coherent, internally consistent and plausible description of a possible future state of the world

science: soft or hard knowledge based on the testing of hypotheses; in *hard* science these hypotheses are approximate descriptions of the real world, and testing is supposedly objective; *soft* science sees them as alternative conceptualizations of something complex, with scientists influenced by the testing: see *GST, SSM* and *social learning*

simulation model model based on mathematical equations and quantitative information, which allows a virtual system to be simulated

social capital *see* capitals

social learning the process, and outcome, of working together to analyse and conceptualise a problem, during which the understandings and practices of the participants change

social-ecological system an area in which human society (*world 3*) interacts with 'nature' (*world 1*); a concrete instance of what is conceptualised in Figure 1.3

SPICOSA Science and Policy Integration for Coastal Systems Assessment, an European research project

SSM Soft System Methods, in which systems are a way of understanding the world; multiple conceptualizations are possible, and observers are not distinct from the system; cf. GST

stakeholder individual, group, or organisation, that has an interest in, or a claim related to, a *human activity* or its *impact*

stakeholder forum a body of stakeholders who meet regularly to exchange views

stakeholder-issue mapping *see* Figure 5.1

STELLA® (www.iseesystems.com/softwares/Education/StellaSoftware.aspx) modelling software with iconic interface, made by isee systems, inc.

sunk costs investments already made, that cannot be reversed

sustainable not depleting *natural capitals* (ecological objective)

system a set of components and relationships within a defined boundary: see Table 3.1; cf. 'the System' of Habermas (see *Lifeworld*)

System Appraisal the fourth step in an application of the *SAF*

System Design the second step in an application of the *SAF*

System Formulation the third step in an application of the *SAF*

System Output the fifth step in an application of the *SAF*

system state see *variable*

timestep repeated small interval of time for which the mathematical equations are solved by the *algorithms* of a *simulation model*

transaction costs costs of participating in a *market*

transformation costs costs of making raw materials into a product

UN United Nations

UWWTD (European) Urban Waste Water Treatment Directive: see Table 1.1

variable: dependent, independent, forcing, or state a dynamic quantity; in a graph, the *dependent variable* (plotted on the vertical or y-axis) result from a transformation of the *independent variable* (plotted on the horizontal or x-axis); the value of a *state variable* helps define *system state*; a *forcing variable* is a quantity supplied to a simulation model and not changed by the results of the simulation – i.e., it is a *boundary condition*

virtual system imagined *system* that approximates all relevant properties of the real system in relation to the *human activities* and *impacts* relevant to the *Issue*

WFD (European) Water Framework Directive: see Table 1.1

widget placeholder name for mechanical or other manufactured device

References

Alexandrov, G.A., Ames, D., Bellocchi, G., Bruen, M., Crout, N., Erechtchoukova, M., Hilde-brandt, A., Hoffman, F., Jackisch, C., Khaiter, P., Mannina, G., Matsunaga, T., Purucker, S.T., Rivington, M. and Samaniego, L. (2011) 'Technical assessment and evaluation of environmental models and software: Letter to the Editor', *Environmental Modelling & Software*, Vol. 26, pp. 328–36.

Anker-Nilssen, T., Høyland, T., Barrett, R., Lorentsen, S.-H. and Strøm, H. (2007) 'Dramatic breeding failure among Norwegian seabirds', *Research News, The Seabird Group Newsletter*, No. 106 (Oct. 2007), pp. 7–8. The Seabird Group, Sandy, Bedfordshire.

Anon (2002) *Review and Synthesis of the Environmental Impacts of Aquaculture*. Edinburgh: Scottish Executive Central Research Unit. Available at: URL: http://www.scotland.gov.uk/Publications/2002/08/15170/9405 (accessed 31 January 2011).

Anon (2004) *World Population to 2300*. New York: United Nations Department of Economic and Social Affairs, Population Division. Available from URL: www.unpopulation.org (accessed 31 January 2011).

Bacher, C. and Gangnery, A. (2006) 'Use of dynamic energy budget and individual based models to simulate the dynamics of cultivated oyster populations', *Journal of Sea Research*, Vol. 56, pp. 14–155.

Barth, F. (1966) *Models of social organisation*, London: Royal Anthropological Institute of Great Britain and Ireland.

Bellinger, G. (2004) 'Systems Thinking: A journey in the realm of systems' (online), http://www.systems-thinking.org/index.htm (accessed 6 February 2011).

Benincà, E., Huisman, J., Heerkloss, R., Jöhnk, K.D., Branco, P., Nes, E.H.V., Scheffer, M. and Ellner, S.P. (2007) 'Chaos in a long-term experiment with a plankton community', *Nature*, Vol. 451, pp. 822–5.

Bergh, J.C.J.M. van den (2001) 'Ecological economics: themes, approaches, and differences with environmental economics', *Regional Environmental Change*, Vol. 2, pp. 13–23.

Bergmann, M., and Schramm, E. (eds) (2008) *Transdisziplinäre Forschung. Integrative Forschungsprozesse verstehen und bewerten*, Frankfurt am Main: Campus Verlag.

Berkes, F. and Folke, C. (eds) (1998) *Linking Social and Ecological Systems. Management Practices and Social Mechanisms for Building Resilience*, Cambridge: Cambridge University Press.

Billen, G., Somville, M., De Becker, E. and Servais, P. (1985) 'A nitrogen budget of the Scheldt hydrographic basin.' *Netherlands Journal of Sea Research*, Vol. 19, pp. 223–30.

Borner, J. (2011) 'Nachhaltigkeitsmanagement: Moderne Kommunikation als Konstruktion von Realität nachhaltiger Entwicklung', in Winzer, P. (ed.) (2011) *Entwicklung im Wuppertaler Generic-Management-Konzept*, Aachen: Shaker Verlag.

Borner, J., Bittencourt, I. and Heiser, A. (eds) (2003) *Nachhaltigkeit in 50 Sekunden*, München: Oekom Verlag.

Bremer, S. (2010) 'Introducing a 'post-normal' science–policy interface for coastal governance according to the principles of Integrated Coastal Zone Management: The SPICOSA Experience'. ms.

Brewer, G.D. (1999) 'The challenges of interdisciplinarity', *Policy Sciences*, Vol. 32, pp. 327–37.

Canu, D.M., Campostrini, P., Dalla Riva S., Pastres, R., Pizzo, L., Rossetto, L., and Solidoro, C. (2011) 'Addressing Sustainability of clam aquaculture in the Venice Lagoon', *Ecology and Society*, special feature on 'A Systems Approach for Sustainable Development in Coastal Zones', in press.

Caroppo, C., Giordano, L., Petrocelli, A., Sclafani, P., Rubino, F., Palmieri, N., Bellio, G., and Bisci, P. (2010) *Output Step Scientific Report SSA 14, Spicosa project report*, Taranto, Napoli: National Research Council.

C.E.C. (1991) 'Council Directive of 21 May 1991 concerning urban waste water treatment (91/271/EEC)', *Official Journal of the European Communities*, Vol. L135 of 30.5.91, pp. 40–52.

Chandler, D. (2001) *Semiotics: the Basics*, London: Routledge. Also available at URL: http://www.aber.ac.uk/media/Documents/S4B/ (visited 20 January 2010).

Checkland, P. (1999) *Systems Thinking, Systems Practice*, Chichester: John Wiley & Sons Ltd.

Checkland, P.B. and Scholes, J. (1990) *Soft Systems Methodology in Action*, Chichester: John Wiley and Sons, Ltd.

Coles, B.J. (2000) 'Doggerland: the cultural dynamics of a shifting coastline' *Geological Society, London, Special Publications*, Vol. 175, pp. 393–401.

Commission of the European Communities (2000) *Communication from the Commission to the Council and the European Parliament on Integrated Coastal Zone Management: a strategy for Europe*, Brussels: COM (2000) 547 final, *pp. 27*.

Common, M. and Stagl, S. (2005) *Ecological Economics*, Cambridge, University Press.

Costanza, R., d'Arges, R., de Groot, R.S., Farber, S., Grasso, M., Hannon, B., Limburg, K., Naeem, S., O'Neill, R.V., Paruelo, J., Raskin, R.G., Sutton, P. and van den Belt, M. (1997) 'The value of the world's ecosystem services and natural capital', *Nature*, Vol. 387, pp. 253–60.

Crossett, K.M. (2005) *Population Trends Along the Coastal United States: 1980–2008*, Washington, D.C.: United States National Oceanic and Atmospheric Administration, Available from URL: http://oceanservice.noaa.gov/programs/mb/supp_cstl_population.html (accessed 4 January 2011).

Crossland, C.J., Kremer, H.H., Lindeboom, H.J., Marshall Crossland, J.I. and Le Tissier, M.D.A., (eds) (2005) *Coastal Fluxes in the Anthropocene: The Land–Ocean Interactions in the Coastal Zone Project of the International Geosphere–Biosphere Programme*, Berlin: Springer-Verlag.

Cuddington, K. (2001) 'The "Balance of Nature" Metaphor and Equilibrium in Population Ecology', *Biology and Philosophy*, Vol. 16, pp. 463–79.

[insert reference:]

Dawkins, R. (1989) *The Extended Phenotype*, Oxford: Oxford University Press.

Delta Works (2004) 'The Flood of 1953' (online), www.deltawerken.com/The-flood-of-1953/89.html (visited 2 January 2011).

Devlin, M., Best, M., Coates, D., Bresnan, E., O'Boyle, S., Park, R., Silke, J., Cusack, C. and Skeats, J. (2007) 'Establishing boundary classes for the classification of UK marine waters using phytoplankton communities', *Marine Pollution Bulletin*, Vol. 55, pp. 91–103.

Diamond, J. (2005) *Collapse: How Societies Choose to Fail or Succeed*, London: Penguin Books.

Dinesen, G., E., Timmermann, K., Roth E., Markager S., Ravn-Jonsen, L., Hjorth, M., Petersen,

J.K., Holmer, M., and Støttrup, J.G. (2011) 'Mussel production and WFD targets in the Limfjord, Denmark: an integrated assessment for use in system-based management', *Ecology and Society*, special feature on 'A Systems Approach for Sustainable Development in Coastal Zones', in press.

Douguet, J.-M., Vanderlinden, J.-P. and Baztan, J. (2011) *KerCoasts, a multimedia tool for learning and deliberation support technology on coastal zone management. A user guide*, Versailles: REEDS, University of Versailles Saint-Quentin-en-Yvelines, *pp.* 50.

Duggins, D.O. (1980) 'Kelp Beds and Sea Otters: an Experimental Approach', *Ecology*, Vol. 61, pp. 447–53.

Eco, Umberto (1976) *A Theory of Semiotics*, Bloomington: Indiana University Press.

Elmgren, R. (2001) 'Understanding human impact on the Baltic ecosystem: Changing views in recent decades', *Ambio*, Vol. 30, pp. 222–31.

Engelen, G. (2004) 'Models in policy formulation and assessment: The WadBOS Decision Support System', *in*: Wainwright, J. and Mulligan, M. (eds) (2004) *Environmental Modelling: Finding simplicity in complexity*, West Sussex: John Wiley and Sons Ltd, pp. 257–71.

EPA (2009) *Guidance on the Development, Evaluation, and Application of Environmental Models*, Washington D.C: Council for Regulatory Environmental Modeling, United States Environmental Protection Agency, *pp.* 90. Available from URL: www.epa.gov/crem/cremlib.html (accessed 7 February 2011).

European Commission (1999) 'Lessons from the European Commission's Demonstration Programme on Integrated Coastal Zone Management (ICZM) – 1997–1999'. Available at URL: ec.europa.eu/environment/iczm/pdf/vol2.pdf (accessed: 7 February 2011).

European Commission (2010) 'Commission Decision of 1 September 2010 on criteria and methodological standards on good environmental status of marine waters', *Official Journal of the European Union*, Vol. L 232, pp. 14–24.

FAO (2009) *The state of world fisheries and aquaculture 2008*, Rome: Food and Agriculture Organization of the United Nations, Fisheries and Aquaculture Department.

Fennel, W. and Neumann, T. (2004) *Introduction to the modeling of marine ecosystems*, Amsterdam: Elsevier.

Finlayson, J.G. (2005) *Habermas, a very short introduction*, Oxford: Oxford University Press.

Frank, K.T., Petrie, B., Choi, J.S. and Leggett, W.C. (2005) 'Trophic Cascades in a Formerly Cod-Dominated Ecosystem', *Science*, Vol. 308, pp. 1621–3.

Franz-Balsen, A. and Heinrichs, H. (2007) 'Managing sustainability communication on campus: experiences from Lüneburg', *International Journal of Sustainability in Higher Education*, Vol. 8, pp. 431–45.

Franzén, F., Elmgren, R., Kratzer, S., Larsson, U., Walve, J., Kinell, G., Söderqvist, T. and Soutukorva, A. (2010) *Output Step Scientific Report SSA 4, Spicosa project report*, Stockholm: Department of Systems Ecology, Stockholm University and Enveco Environmental Economics Consultancy Ltd.

Franzén, F., Kinell, G., Walve, J., Söderqvist, T. and Elmgren, R. (2011) 'Stakeholder Involvement in Adaptive Management of Coastal Eutrophication in Himmerfjärden, Baltic Coast of Sweden', *Ecology and Society*, special feature on 'A Systems Approach for Sustainable Development in Coastal Zones', in press.

Frigg, R. and Hartmann, S. (2006) 'Models in Science' (online), URL:http//plato.stanford.edu/entries/models-science/ (accessed 6 May 2010).

Frodeman, R., Klein, J. T. and Mitcham, K. (eds) (2010) *The Oxford Handbook of Interdisciplinarity*, Oxford: Oxford University Press.

German Ministry for Economic Cooperation and Development (2011) 'Glossary: ownership', www.bmz.de/de/service/glossar/O/ownership.html (visited 10 January 2011).

Gersman, H. (2010) *Klimakommunikation in Deutschland* (DVD), Berlin–Santiago: KMGNE, Universidad Internacional.

Gilbert, A., Bacher, C., d'Hernoncourt, J., Fernandes, T., Konstantinou, Z., Lowe, C., McFadden, L., Mette, A., Priest, S., Raux, P., Salomons, W., Tett, P., Tunstall, S. and Stojanovic, T. (2011), *Coastal SAF Glossary, Spicosa project report*, Amsterdam: VU University, available at: www.coastal-saf.eu.

Gladwell, M. (2002) *The Tipping Point: how little things can make a big difference*, London: Abacus.

Golüke, U. (2001) *Making Use of the Future – Scenario Building as a Tool to Solve Regional Autonomy Conflicts*, Working-Paper, München: Centrum für angewandte Politikforschung.

Government of Norway (2006) *St.prp. no 32 (2006–2007) About protection of wild salmon and establishment of national salmon rivers and national salmon fjords*, Oslo: Government Printer.

Grimm, V. and Railsback, S.F. (2005) *Individual-based Modeling and Ecology*, Princeton, New Jersey: Princeton University Press.

Grimnes, A. and Jakobsen, P.J. (1996) 'The physiological effects of salmon lice infection on post-smolt of Atlantic salmon', *Journal of Fish Biology*, Vol. 46, No. 6, pp 1179–1194.

Guggenheim, D. (Director) (2006) *An Inconvenient Truth: A Global Warning*, Hollywood: Paramount.

Gunderson, L.H., and Holling, C.S. (eds) (2002) *Panarchy: understanding transformations in human and natural systems. Washington DC:* Island Press.

Habermas, J. (1981). *Theorie des kommunikativen Handelns 1, Handlungsrationalität und gesellschaftliche Rationalisierung*, Frankfurt am Main: Suhrkamp. Published in English in 1984, as *The Theory of Communicative Action. Volume 1: Reason and the Rationalization of Society*, Boston, MA: Beacon Press.

Habermas, J. (1992) 'Drei normative Modelle der Demokratie: Zum Begriff deliberativer Demokratie', in Münkler, H. (ed.) *Die Chancen der Freiheit. Grundprobleme der Demokratie*, München und Zürich: Piper Verlag.

Hails, C., Humphrey, S., Loh, J. and Goldfinger, S., (eds) (2008), *Living Planet Report 2008*, Gland, Switzerland: WWF-World Wide Fund for Nature. Available from URL: www.panda.org/about_our_earth/all_publications/living_planet_report/ (accessed 7 February 2011).

Hardin, G. (1968) 'The Tragedy of the Commons', *Science*, Vol. 162, pp. 1243–48.

Heemskerk, M., Wilson, K. and Pavao-Zuckerman, M. (2003) 'Conceptual models as tools for communication across disciplines', *Conservation Ecology*, Vol. 7, No. 8, http://www.consecol.org/vol7/iss3/art8.

Hobbs, F. and Stoops, N. (2002) *Demographic Trends in the 20th Century. Census 2000 Special Reports, Series CENSR-4*, Washington D.C: (U.S. Government Printing Office): U.S. Census Bureau.

Holland, M. (1999) *Pressure Points, The theory and practice of communications planning for planners in sustainable development projects*, MSc thesis, Canada: The University of British Columbia.

Hopkins, T.S., Bailly, D. and Engelen, G. (2011) 'The Systems Approach adapted to Coastal Zones', *Ecology & Society*, special feature on 'A Systems Approach for Sustainable Development in Coastal Zones', in press.

Hughes, D. and Nickell, T. (2009) *Recovering Scotland's Marine Environment: Report to*

Scottish Environment LINK, Oban, Scotland: Scottish Association for Marine Science Internal Report no. 262, *pp.* 63, available from URL: http://www.sams.ac.uk/research/departments/ecology/ecology-projects/.

Involve (2005) *People & participation – how to put citizens at the heart of decision-making*, Boston, MA: Beacon Press. Also available at URL: www.involve.org.uk/assets/Uploads/People-and-Participation.pdf (accessed 3 January 2011).

IPCC (2010) *Intergovernmental Panel on Climate Change*, www.ipcc.ch/organization/organization.shtml (accessed 7 February 2011).

IPCC-TGICA (2007) *General Guidelines on the Use of Scenario Data for Climate Impact and Adaptation Assessment. Version 2.* Prepared by T.R. Carter on behalf of the Intergovernmental Panel on Climate Change, Task Group on Data and Scenario Support for Impact and Climate Assessment, Helsinki: Finish Environmental Institute.

Ives, A.R. and Carpenter, S.R. (2007) 'Stability and Diversity of Ecosystems', *Science*, Vol. 317, pp. 58–62.

Jackson, J.B.C., Kirby, M.X., Berger, W.H., Bjorndal, K.A., Botsford, L.W., Bourque, B.J., Bradbury, R.H., Cooke, R., Erlandson, J., Estes, J.A., Hughes, T.P., Kidwell, S., Lange, C.B., Lenihan, H.S., Pandolfi, J.M., Peterson, C.H., Steneck, R.S., Tegner, M.J. and Warner, R.R. (2001) 'Historical Overfishing and the Recent Collapse of Coastal Ecosystems', *Science*, Vol. 293, pp. 629–37.

Jakobson, R.O. (1971) *Selected Writings*, Vol. 2, Ruby, S. (ed.), Mouton: The Hague.

Jennings, S. and Polunin, N.V.C. (1996) 'Impacts of Fishing on Tropical Reef Ecosystems', *Ambio*, Vol. 25, pp. 44–49.

Johnson, L.E. (1991) *A Morally Deep World*, Cambridge: Cambridge University Press.

Klewes, J. and Langen, R. (eds) (2008) *Change 2.0, Beyond Organisational Transformation*, Berlin: Springer Verlag.

Krebs, C.J. (1988) *The Message of Ecology*, New York: Harper & Row.

Kuhn, T.S. (1996) *The Structure of Scientific Revolutions*, Chicago: University of Chicago Press.

Kurlansky, M. (1998) *Cod; a biography of the fish that changed the world*, London: Jonathan Cape.

Laurent, C., Tett, P., Fernandes, T., Gilpin, L. and Jones, K.J. (2006) 'A dynamic CSTT model for the effects of added nutrients in Loch Creran, a shallow fjord', *Journal of Marine Systems*, Vol. 61, pp. 149–64.

Lemons, J. (ed.) (1996) *Scientific uncertainty and environmental problem solving*, Cambridge, MA: Blackwell Science.

Lindahl, O., Hart, R., Hernroth, B., Kollberg, S., Loo, L.-O., Olrog, L., Ann-Sofi Rehnstam-Holm, Svensson, J., Svensson, S. and Syversen, U. (2005) 'Improving Marine Water Quality by Mussel Farming: A Profitable Solution for Swedish Society', *Ambio*, Vol. 34, pp.131–8.

Link, J.S., Bogstad, B., Sparholt, H. and Lilly, G.R. (2009) 'Trophic role of Atlantic cod in the ecosystem', *Fish and Fisheries*, Vol. 10, pp. 58–87.

Lonergan, S. (2004) *Introduction in Understanding Environment, Conflict and Cooperation*, Paris: United Nations Environmental Programme.

Luhmann, N. (1985) *A sociological theory of law*, London: Routledge and Kegan Paul.

Luhmann, N. (1989) *Ecological Communication* (translated by Bednarz, John), Chicago: University of Chicago Press.

Luiten, H. (1999) 'A legislative view on science and predictive models', *Environmental Pollution*, Vol. 100, pp. 5–11.

Lull, J. (1995) *Media, Communication, Culture*, New York: Columbia University Press.

Macaulay, V., Hill, C., Achilli, A., Rengo, C., Clarke, D., Meehan, W., Blackburn, J., Semino, O., Scozzari, R., Cruciani, F., Taha, A., Shaari, N.K., Raja, J.M., Ismail, P., Zainuddin, Z., Goodwin, W., Bulbeck, D., Bandelt, H.-J., Oppenheimer, S., Torroni, A. and Richards, M. (2005) 'Single, Rapid Coastal Settlement of Asia Revealed by Analysis of Complete Mito-chondrial Genomes', *Science*, Vol. 308, pp. 1034–6.

MacKenzie, B.R., Bager, M., Ojaveer, H., Awebro, K., Heino, U., Holm, P. and Must, A. (2007) 'Multi-decadal scale variability in the eastern Baltic cod fishery 1550–1860 – evidence and causes,' *Fisheries Research*, Vol. 87, pp. 106–19.

Marx, K. (1977) *Preface to: A Contribution to the Critique of Political Economy (originally published in 1859)*, Moscow: Progress Publishers. Available at: http://www.marxists.org/archive/marx/works/1859/critique-pol-economy/preface.htm

Maurstad, A. (2004) 'Cultural seascapes : preserving local fishermen's knowledge in Northern Norway', in Krupnik, I., Mason, R. and Horton, T. W. (ed.) (2004) *Northern ethnographic landscapes: perspectives from circumpolar nations*, Washington, D.C: Arctic Studies Center, National Museum of Natural History, Smithsonian Institution in collaboration with the National Park Service.

McAnany, P.A. and Yoffee, N. (eds) (2009). *Questioning Collapse: Human Resilience, Ecological Vulnerability, and the Aftermath of Empire,* Cambridge: Cambridge University Press.

McCann, K. S. (2002) 'The diversity–stability debate', *Nature*, Vol. 405, pp. 228–33.

McFadden, L. and Priest, S. (2011) *Guidance for identifying the Policy Issue, Spicosa Project Report*, London: Flood Hazard Research Centre, Middlesex University.

McFadden, L., Priest, S. and Green, C. (2010) *Introducing institutional mapping: a guide for SPICOSA scientists, Spicosa Project Report,* London, Flood Hazard Research Centre, Middlesex University.

McFadden, L., Priest, S., Green, C. and Sandberg, A. (2009) *Science and Policy Integration, Spicosa Project Report*, London: Flood Hazard Research Centre, Middlesex University.

McGuire, C. and McGuire, C. (2004) 'Ecofeminist Visions' in Dolbeare, K. and Cummings, M.S. (eds) (2004) *American Political Thought, 5th edition,* Washington D.C: Congressional Quarterly Press, pp. 512–17.

Megapesca (1997) 'Latvian fisheries – ready for exports' (online), URL:http://www.gisl.co.uk/megapesca/latviawebsite.html (accessed 1 January 2011).

Melaku Canu, D., Campostrini, P., Dalla Riva, S., Pastres, R., Pizzo, L., Rossetto, L. and Solidoro, C. (2011) 'Addressing sustainability of clam farming in the Venice Lagoon', *Ecology and Society*, special feature on 'A Systems Approach for Sustainable Development in Coastal Zones,' in press.

Mellars, P. (2006) 'Why did modern human populations disperse from Africa ca. 60,000 years ago? A new model', *Proceedings of the National Academy of Science of the United States of America*, Vol. 103, pp. 9381–6.

Melton, N.D. and Nicholson, R.A. (2004) 'The Mesolithic in the Northern Isles: the prelimi-nary evaluation of an oyster midden at West Voe, Sumburgh, Shetland, U.K', *Antiquity*, Vol. 78, no. 299. Available at: antiquity.ac.uk/ProjGall/nicholson/ (accessed 7 February 2011).

Merton, R.K. (1968) *Social Theory and Social Structure*, New York: Free Press.

Mette, A. *et al.* (2011) *The Systems Approach Framework Handbook*, www.coastal-saf.eu

Mette, A., Borner, J., Sandberg, A., Vanderlinden, J-P., Priest, S., McFadden, L., Lowe, C., D'Hernoncourt, J., Fernandes, T., Hirschfeld, J. and Raux, P. (2011) *Guide to System Output, SPICOSA Project Report*, Berlin, KMGNE.

References

Michelsen, G. and Godemann, J., (eds) (2007) *Handbuch der Nachhaltigkeitskommunikation, Grundlagen und Praxis*, Muenchen, Oekom Verlag.

Millennium Ecosystem Assessment (2003) *Ecosystems and Human Well-Being: A Framework for Assessment*, Washington D.C: Island Press.

Millennium Ecosystem Assessment (2005) *Ecosystems and Human Well-being: Synthesis*, Washington, D.C.: Island Press.

Mithen, S. (2003) *After the Ice*, London, Weidenfeld & Nicolson.

Morecroft, J. (2007) *Strategic Modelling and Business Dynamics: a feedback systems approach*, Chichester: John Wiley and Sons Ltd.

Morgan, E. (1982) *The aquatic ape: a theory of human evolution*, London: Souvenir Press.

Moss, B. (2008) 'The Water Framework Directive: Total environment or political compromise?' *Science of the Total Environment*, Vol. 400, pp. 32–42.

Naylor, R. and Burke, M. (2005) 'Aquaculture and ocean resources: raising tigers of the sea', *Annual Review of Environment and Resources*, Vol. 30, pp. 185–218.

Nellemann, C. and Corcoran, E. (ed.) (2010) *Dead planet, living planet: Biodiversity and ecosystem restoration for sustainable development. A Rapid Response Assessment.* United Nations Environment Programme. Arendal: GRID-Arendal. Obtainable from URL: www.grida.no/publications/rr/dead-planet/ (accessed February 4, 2011).

Nicholls, R.J. and Small, C. (2002) 'Improved estimates of coastal population and exposure to hazards released' *EOS Transactions of the American Geophysical Union*, Vol. 83, pp. 301–5.

Nimmo, F., Capell, R., Huntington, T., Grant, A. and Leach, J. (2009) 'Assessment of Evidence that Fish Farming Impacts on Tourism (SARF045)', *Scottish Aquacultural Research Forum Report*, available from URL:www.sarf.org.uk.

Nizzoli, D., Bartoli, M., and Viaroli, P. (2006) 'Nitrogen and phosphorous budgets during a farming cycle of the Manila clam *Ruditapes philippinarum*: An in situ experiment', *Aquaculture*, Vo. 261, pp. 98–108.

North, D.C. (1990) *Institutions, institutional change and economic performance, The Political economy of institutions and decisions*, New York: Cambridge University Press.

North, W.J. and Pearse, J.S. (1970) 'Sea urchin population ecplosions in southern Californian coastal waters', *Science*, Vol. 167, pp. 209.

Odum, E.P. (1959) *Fundamentals of Ecology*, Philadelphia: Saunders.

Odum, H.T. and Odum, E.C. (2000) *Modelling for all scales: an introduction to system simulation*, San Diego: Academic Press.

Oppenheimer, S. (2004) *Out of Eden – The peopling of the world*, London: Constable and Robinson.

OSPAR (2003) *2003 Strategies of the OSPAR Commission for the Protection of the Marine Environment of the North-East Atlantic (Reference number: 2003–21)*, Bremen: Ministerial Meeting of the OSPAR Commission.

Ostrom, E. (1990) *Governing the Commons: The evolution of Institutions for Collective Action*, Cambridge: Cambridge University Press.

Ostrom, E. (2005) *Understanding Institutional Diversity*, New York: Princeton University Press.

Ostrom, E. (2007) 'A diagnostic approach for going beyond panaceas', *Proceedings of the National Academy of Science of the United States of America*, Vol. 104, pp. 15181–7.

Pereira, Guimarães Â. and Brilhante, De Sousa P. (eds) (2005) *Knowledge Assessment Methodologies Fall School Manual*, PB/2005/IPSC/0384, Ispra: European Commission Joint Research Centre, Institute for the Protection and the Security of the Citizen.

References

Pinnegar, J.K., Polunin, N.V.C., Francour, P., Badalamenti, F., Chemello, R., Hrmelin-Vivien, M.-L., Hereu, B., Milazzo, M., Zabala, M., D'Anna, G. and Pipitone, C. (2000) 'Trophic cascades in benthic marine ecosystems: lessons for fisheries and protected-area management', *Environmental Conservation*, Vol. 27, pp. 179–200.

Poole, W. (2010) *The world makers: scientists of the restoration and the search for the origin of the earth, the past in the present*, London: Oxford University Press, Peter Lang.

Popper, K. (1978) *Three worlds. The Tanner Lecture on Human Values, delivered at the University of Michigan, April 7, 1978*. Salt Lake City: University of Utah. Obtainable at URL: www.tannerlectures.utah.edu/lectures/documents/popper80.pdf (visited 23 January 2011).

Popper, K. (2002). *The Poverty of Historicism* (2nd edition; 1st English language edition in 1957; in German in 1936), London: Routledge.

Popper, K.R. (2002) *The logic of scientific discovery*, London: Routledge Classics.

Portilla, E., Tett, P., Gillibrand, P. A. and Inall, M. (2009) 'Description and sensitivity analysis for the LESV model: water quality variables and the balance of organisms in a fjordic region of restricted exchange.' *Ecological Modelling*, Vol. 220, pp. 2187–205.

Poteete, A.R., Janssen, M.A. and Ostrom, E. (2010) *Working together: collective action, the commons, and multiple methods in practice*, Princteon N.J: Princeton University Press, Princeton, N.J.

Press, W.H., Flannery, B.P., Teukolsky, S.A. and Vetterling, W.T. (1989) *Numerical Recipes in Pascal*, New York, Cambridge University Press.

Refsgaard, J.C. and Henriksen, H.J. (2004) 'Modelling guidelines – terminology and guiding principles', *Advances in Water Resources*, Vol. 27, pp. 71–82.

Rolston III, H. (1994) 'Value in Nature and the Nature of Value', *in*: Attfield, R. and Belsey, A. (eds) *Philosophy and the Natural Environment*, Cambridge: Cambridge University Press.

Rousseau, J.J. (1754) *Discourse on the Origin and Basis of Inequality Among Men*, Amsterdam: Academy of Dijon.

Rykiel, E.J. (1996) 'Testing ecological models: the meaning of validation', *Ecological Modelling*, Vol. 90, pp. 229–44.

Sala, E., Boudouresque, C. F. and Harmelin-Vivien, M. (1998) 'Fishing, Trophic Cascades, and the Structure of Algal Assemblages: Evaluation of an Old but Untested Paradigm', *Oikos*, Vol. 82, pp. 425–39.

Sandberg, A. (2010) 'Institutional challenges to robustness of delta and floodplain agricultural systems', *Environmental Hazards*, Vol. 9, pp. 284–300.

Sastre, S., Tomlinson, B., and Blasco, D. (2010) *Output Step Scientific Report SSA 12, Spicosa project report*, Barcelona: CSIC.

Scheffer, M. and Carpenter, S.R. (2003) 'Catastrophic regime shifts in ecosystems: linking theory to observations', *TRENDS in Ecology and Evolution*, Vol. 18, pp. 648–56.

Scheffer, M., Bascompte, J., Brock, W.A., Brovkin, V., Carpenter, S.R., Dakos, V., Held, H., Nes, E.H.V., Rietkerk, M. and Sugihara, G. (2009) 'Early-warning signals for critical transitions', *Nature*, Vol. 461, pp. 53–9.

Scheffer, M., Carpenter, S., Foley, J.A., Folke, C. and Walker, B. (2001) 'Catastrophic shifts in ecosystems', *Nature*, Vol. 413, pp. 591–6.

Schumpeter, J.A. (1934) *The theory of economic development: an inquiry into profits, capital, credit, interest, and the business cycle*, Cambridge, Mass: Harvard University Press.

Scott, N.J. and Parsons, E.C.M. (2005) 'A survey of public opinion in south-west Scotland on cetacean conservation issues' *Aquatic Conservation: Marine and Freshwater Ecosystems*, Vol. 15, pp. 299–312.

References

Sharma, S.K. and Starik, M. (eds) (2004) *Stakeholders, the Environment and Society*, Cheltenham: Edward Elgar Publishing.

Shipman, B. and Stojanovic, T. (2007) 'Facts, Fictions, and Failures of Integrated Coastal Zone Management in Europe', *Coastal Management*, Vol. 35, pp. 375–98.

Smith, D.E. (1999) *Writing the social: critique, theory, and investigations*, Toronto: University of Toronto Press.

Snow, C.P. (1993) *The two cultures and the Scientific Revolution*, Cambridge: Cambridge University Press.

Soetaert, K. and Herman, P.M.J. (2009) *A Practical Guide to Ecological Modelling*, Springer.

Stal, L.J., Albertano, P., Bergman, B., von Brockel, K., Gallon, J.R., Hayes, P.K., Sivonen, K. and Walsby, A.E. (2003) 'BASIC: Baltic Sea cyanobacteria. An investigation of the structure and dynamics of water blooms of cyanobacteria in the Baltic Sea – responses to a changing environment', *Continental Shelf Research*, Vol. 23, pp. 1695–1714.

Stern, N. (2006) *Stern Review, The Economics of Climate Change*, London: H.M. Treasury.

Stiglitz, J. (2010) *Freefall: Free Markets and the Sinking of the Global Economy*, New York: W.W. Norton & Company, Inc.

Stinchcombe, A. (1968) *Constructing Social Theories*, New York: Harcourt, Brace & World, Inc.

Stockholm Resilience Centre (2010) Resilience Dictionary. URL: http://www.stockholmresilience.org/research/whatisresilience/resiliencedictionary.4.aeea469 11a3127427980004355.html (visited 29 December 2010).

STOWA/RIZA (1999) *Smooth modelling in Water management, Good Modelling Practice Handbook; STOWA report 99-05*, Dutch Deptartment Of Public Works, Institute for Inland Water Management and Waste Water treatment report 99.036.

SustainAbility Ltd. (ed) (1996) *Engaging Stakeholders*. London: SustainAbility Ltd.

Suttle, C.A. (2000) 'Cyanophages and their role in the ecology of cyanobacteria'. in Whitton, B. A. and Potts, M. (ed.) (2000) *The Ecology of Cyanobacteria: Their Diversity in Time and Space*, Dordrecht: Kuwer Acadenic Publishers.

Tett, P. (2008) 'Fishfarm wastes in the ecosystem', in Holmer, M., Black, K., Duarte, C.M., Marbá, N. and Karakassis, I., (eds) (2008) *Aquaculture in the Ecosystem*, Berlin: Springer, pp. 1–46.

Tett, P., Mongruel, R., Levrel, H., Hopkins, T., Sandberg, A., Hadley, D., Fernandes, T., Hendrick, V., Mette, A., Vermaat, J., Gilbert, A., d'Hernoncourt, J., McFadden, L., Priest, S., Green, C., and d'Alconà, M.R. (2011) *Guide to System Design, v.3.09, SPICOSA Project Report*, Oban, Scottish Association for Marine Science.

Tett, P. and Wallis, A.C. (1978), 'The general annual cycle of chlorophyll in Loch Creran', Journal of Ecology, Vol. 26, pp.227-239. Thurstan, R.H., Brockington, S. and Roberts, C.M. (2010) 'The effects of 118 years of industrial fishing on UK bottom trawl fisheries', *Nature Communications*, DOI: 10.1038/ncomms1013.

Tolun, L.G., Ergenekon, S., Murat-Hocaoğlu, S., Tülay Çokacar, T., Aslı S., Dönerta , A.S., Hüsrevoğlu, Y.S., Polat-Beken, C., Avaz, G., and Baban, A. (2011) 'Socio-economic Response to Water Quality: A First Experience in Science and Policy Interaction for the Izmit Bay Coastal System', *Ecology and Society*, special feature on 'A Systems Approach for Sustainable Development in Coastal Zones', in press.

Vanderlinden, J-P., Stojanovic, T., Schmuëli, D., Bremer, S., Kostrzewa, C. and McFadden, L. (with others) (2011) *The SPICOSA Stakeholder-Policy Mapping Users' Manual*, Spicosa Project Report, Guyancourt, Paris: Université de Versailles-Saint-Quentin-en-Yvelines.

Vanderlinden, J.-P. (2011) *D1.8 Updated review of the deliberation tool: lessons learned from its use in a coastal context*, SPICOSA, Guyancourt, Paris: Université de

Versailles-Saint-Quentin-en-Yvelines.

Vermaat, J.E., McQuatters-Gollop, A., Eleveld, M.A. and Gilbert, A.J. (2008) 'Past, present and future nutrient loads of the North Sea: Causes and consequences', *Estuarine Coastal and Shelf Science*, Vol. 80, pp. 53–9.

von Bertalanffy, L. (1968) *General Systems Theory: Foundations, Development, Applications*, New York: George Braziller.

Weingart, P. (2001) *Die Stunde der Wahrheit? Zum Verhältnis der Wissenschaft zu Politik, Wirtschaft und Medien in der Wissensgesellschaft*, Weilerswist: Velbrück Wissenschaft.

Weninger, B., Schulting, R., Bradtmöller, M., Clare, L., Collard, M., Edinborough, K., Hilpert, J., Jöris, O., Niekus, M., Rohling, E.J. and Wagner, B. (2008) 'The catastrophic final flooding of Doggerland by the Storegga Slide tsunami', *Documenta Praehistorica*, Vol. 35, pp. 1–24.

Wilkinson, R. and Pickett, K. (2009) *The Spirit Level: Why More Equal Societies Almost Always Do Better*, London, Allen Lane.

Woodcock, G.L. (1986) *Planning, Politics and Communications, Aldershot:* Gower Publishing Company Ltd.

Young, N. and Matthews, R. (2010) *The aquaculture controversy in Canada: activism, policy, and contested science*, Vancouver: UBC Press.

Index

Page numbers in *italic* denote figures. Page numbers in **bold** denote tables.

acqua alta, Venice 81, 101
Actors 20, 28, 45, 49, 115-116, **119**, 134
Adapting Mosaic scenario **72**, *74*
adaptive capacity 144, 148-149
algae, blooms 32, 61, 76
algorithms 90-92, 96-97, 102
Analytical Approach 75
aquaculture 6, 9, 25
    effect on marine ecosystem 33
    shellfish
        Lagoon of Venice 38-39, 82-83
            conceptual model 84-89
            institutional map 49, 99
Asia, coastal zone vulnerability 145
Australia
    coastal zone vulnerability 145
    Oceans Policy 16
autarky 144-145

baccalao 33
Baltic Sea
    algal blooms 32, 61
    fisheries and market arrangements 12-13
Basin de Marennes Oléron, oyster farming,
    visualisation 129-130, *131*, *132*
biodiversity
    Convention on Biological Diversity 15
    in ecosystems 62, 76
black box systems 58, 62, 69-71, *74*, 75, 127
'boom and bust' 61
boundaries 53-54, 57
    coastal zone systems *55*
boundary conditions 57, 70, 97-98, 151
Brahmaputra Delta 145
Buenos Aires, coastal population 3

capital
    durable 67
    financial 67
    human 67
    natural 26, 67
    social 7, 67
carrying capacity 64
catch-quota systems 33
CATWOE **119**, 134
change management 109
chaotic behaviour 62, 63
China, Marine Environmental Protection
    and Prevention and Control of Water
    Pollution Laws 16
chlorophyll 66
    Loch Creran, simulation testing 96-98
cities, coastal zone 1-4
clam fisheries
    STELLA modelling subprogram 92-94
    Venice Lagoon 38-39, 82-83
        conceptual model 84-89
        institutional map 49, 99
climate change mitigation
    Marshland Protection Policy,
        Mecklenburg-Vorpommern 127
    visualisation 129
coastal systems, modelling 79-101, 138-139
coastal zone
    border zone 29-30, 137
        fauna 29-30
    boundaries *55*
    dynamics 31-32
    food 3
    human-induced stress 137-138
    modern cities 2-4, 137

prehistory 3, 25
problems 29-51
  human damage to ecosystems 36-37
  identification 48-50
  nature causes problems for
    communities 37
  representation 36-40
  social system conflict 37
value to humans 2, 3, 6-7, 25
cod
  farmed 33
  wild 33
    overfishing 62
codes 109, 134
Cogito ergo sum 55
collective arrrangements 12
communication
  science and society 103-105
    change management 109
    cultural differences 105-107
    decision-making support tools 114
    deliberation support tools 114-117
    dialogue 107-110
    participation tools 112-113
    risk 110
    scientific information 109
    sustainability 108-110
    visualisation 129-130, *131*, *132*
communications space *23*, 50
computers, modelling programs 89-98,
  101-102
conceptual models 22, 57
  clam fisheries 84-89
  design 125
  freshwater flow, Charente estuary *131*
  use by stakeholders 114
conservation stage, panarchy 64
Convention on Biological Diversity 15
coral reefs
  damage by fishing 32
  protection 145
  value to humans 6-7
costs
  sunk 39-40, 156
  transaction 32, 156
  transformation 32, 156
culture, differences, science and society

105-107
Customers **119**, 134

decision-making, support tools 114
deliberation 18
  matrix 115-116
  support tools 114-117
denitrification 140
dependent variable 58
Descartes, René
  Cogito ergo sum 55
  Discourse on Method (1637) 139
dialogue, science-society 107-110, 116
Diamond, Jared, *Collapse* (2005) 20, 27
diversity 62
Dogger Bank, Stone Age inhabitation 4, 25
DPSIR (Driver, Pressure State, Impact
  Response) 48
drainage *see* land reclamation
dynamic energy budget 85-86

Easter Island, human settlement 20
ecology 13
economics 13
Ecosystem Services for Poverty Alleviation
  146
Ecosystem-Based Management 146
ecosystems 10, 152
  collapse *46*, 62
  damage by humans 36
  damage to humans 37
  goods and services 11
  influence of social and economic
    dynamics 32
  modification by humans 39
  panarchy 64-65
  self-regulation 20
efficiency, economic 13, 67, 152
emergent properties 19, 59, 152
energy use 8
Environment
  CATWOE tool **119**, 134
  legislation 15-17
equity, socio-ecological systems 13, 67, 152
European Union
  Coastal Policy 48
  Common Fisheries Policy 13

Directives, coastal zone 15, **16**
Marine Strategy Framework Directive **16**,
  48, 66, 145
    variables 66
  Maritime Spatial Planning Policy 48
  Nitrates Directive **16**, 140
  Urban Waste Water Treatment Directive
    16, 70, 140
  Water Framework Directive 16, 33, 34,
    48, 139-140
      management 68
      programmes of measures 68, 155
      variables 66-67
eutrophication 152
  Himmerfjärden, Sweden 47, 48, 100
    institutional map 118, *120*
    issue identification 121, **122**
  Limfjorden mussel fisheries 38, 140
  scenario 69
evolutioin 60-61
exploitation stage, panarchy 64
ExtendSim computer program 92, 102
  testing 96-98
externalities 68, 152

fauna, coastal zone border 29-30
feedback loops 58-61
  negative 19, 38, *55*, 59, *60*, 61, 152-153
    Venetian clam fisheries modelling
      88-89
  positive 19, *55*, 59, 61, 152-153
    Venetian clam fisheries modelling 89
fish farming 6, 9
  effect on marine ecosystem 33
  pollution 9, 33, 38
  sea-lice parasites 35, 39
  and wild salmon 9, 35
fisheries 6, 25
  catch-quota systems 33
  collective arrrangements 12
  market arrangements 12-13
fishing
  drift-net, effect on wild salmon 35
  industrial 6
flooding
  Lagoon of Venice 81
  Netherlands 4

food, coastal zone 3
footprint, ecological 8, 26, 152
forcing variable *see* independent variable
forums 114, 134
fossil fuel 8
functional alternatives 61-62

Gas Laws 66, 77
gas platforms 6
General Systems Theory (GST) 27, 44, 53,
  54, 57, 65, 153
  and Soft Systems Methodology 138
glaciation, sea level change 4
Global Orchestration scenario **72**, *74*
Gore, Al, *An Inconvenient Truth* (2006),
  visualisation 129
governance 13-17, 26-27, 143, 153
  collective level 15, 68
  constitutional level 15, 68
  European Union Water Framework
    Directive 68
  functions and levels 15
  operational level 15, 68
Gross Domestic Product 77
Gross Global Product 7
gubernator 14

Habermas, Jürgen
  communicative rationality 18, 27, 111,
    112
  lifeworld 27, 108, 112, 143, 153-154
hard systems 55, 57, 65, 133
heirarchy 54, 88
hierarchical arrangements 12
Himmerfjärden, Sweden
  nitrogen runoff 47, 48, 100
    institutional map 118, *120*
    issue identification 121, **122**
homeostasis 19-20, 60, 153
Homo sapiens 2
  littoralis 3, 24, 137
  societatis 7, 13-14, 26-27, 138
  see also humans
humans
  attraction to coastal zone 2-3, 25, 137
  energy and resource use 8-9, 26
  and environment 10-13

population growth 8, 24
    prehstory 2, 24
    as systems 19-20
hurricane Katrina 7, 25
hydro-electric power stations, effect on wild
    salmon 35

icons, computer models 92, 98
independent variable 58
individual based model 86
input-output relationship 58-59, 62-63,
    69-71, *74*, 75, 127
Institutional Analysis 45
institutional mapping 49
    Lagoon of Venice acquaculture *99*
    SAF application 118, *120*
institutions 14, 49, 146, 153
Integrated Coastal Zone Management
    (ICZM) 34, 140-141, 142
interdisciplinarity 44, 142
Intergovernmental Panel on Climate
    Change (IPCC) 43
    Copenhagen conference 2009 14
    scenarios 70, **71**, 77
interpretative analysis 126-127
Issue 21, *22*, 50, 153
Issue Identification 21, *22*, 118, 120-125
    eutrophication, Himmerfjärden **120**, 121
    mussel farming, Mar Piccolo, Taranto
        121-125
    stakeholders 50
Istanbul, coastal population 3

Jakarta, coastal population 3
jetsam, plastic 6

K-selected species 64
κυβερνήτης 14

Lagos, coastal populatioin 3
land reclamation, Netherlands 4-5
Latvia, fisheries 13
legislation 15-17
lice *see* salmon lice
lifeworld 27, 108, 112, 143, 153-154
Limfjord, Denmark, mussel fishery 33, 38,
    140

linear relationship 59
Loch Creran, phytoplankton chlorophyll,
    simulation testing 96-98
Loch Fyne, SPICOSA SAF application 49
logical structures 60, 76
logistic curve 59
LOICZ (Land Ocean Interactions in the
    Coastal Zone) 2, 15, 146
London
    coastal population 3
    waste disposal 3-4, 25
Luhmann, Niklas, *Ecological Communication*
    (1989) 108
Luther, Martin, vernacular Bible 107

Mar Piccolo, mussel farming 121-125
markets 12, 154
Marshland Protection Policy, Mecklenburg-
    Vorpommern, climate change mitigation
    127
Matlab 101-102
meetings, public 105-106
Mekong Delta 145
Millennium Ecosystem Assessment 67, 68,
    72-75, 77, 146
    scenarios 72-75, **72**
    valuation tools 73, 75
modelling, coastal systems 79-101
models 79
    computer programs 89-98, 101-102
    conceptual 22, 57, 84-89, 114, 125
    icon-based software 92, 98
    individual based 86
    simulation 22, 89-95, 102, 125, 139, 155
        testing 96-97, 102
        testing results 95-98, 102
multidisciplinarity 126, 135
Mumbai, coastal population 3
mussel farming
    Limfjord, Denmark 33, 38
    Mar Piccolo, Taranto 121-125
    scenario 69-70, 71

nature, total global value 7
Netherlands
    flooding 4
    land reclamation 4-5, 25

New Orleans, hurricane Katrina, population 7, 25
New York, coastal population 2-3
NOAA (National Oceanic and Atmospheric Administration) 2
North America, coastal zone vulnerability 145
North Atlantic Ocean, wild cod 33
North Sea, modern environment 5-6, 25

Oder Estuary marshlands 127
oil platforms 6
Order from Strength scenario **72**, *74*
organic matter *see* particulate organic matter
orrery 79, *80*
OSPAR Strategy to Combat Eutrophication 140
Output Step 127-129
overfishing 32, 62
Owners, CATWOE tool **119**, 134
ownership 103, 133-134
oyster farming, Pertuis Charentais 129-130

panarchy 38, 64-65, 76
paradigms, incompatibility between disciplines 44
parameters 91, 98
participation tools 112-113
particulate organic matter 83, 85
Pertuis Charentais, oyster farming, visualisation 129-130, *131, 132*
phenomena
    information world 56
    mental world 56
    physical world 56
phytoplankton 66
    chlorophyll, Loch Creran, simulation testing 96-98
    and shellfish aquaculture 83
        conceptual model 84-89
    STELLA modelling subprogram 92-94
plastic waste 6
policy 31
    instruments 31
    problems, coastal zone 29-51

policy-issue mapping 47, 154
policy-making 40-48
    challenges 41-42
    role of science 42-45
policy-stakeholder mapping 47, 154
Popper, Karl 'three worlds' 56, 143
population
    growth 8, 24
    living in coastal zone 2-3, 24
problem scaling 125
problems
    coastal zone, representation 36-40
    crossover 37, 44
property rights 14
puffins *30*

r-selected species 64
rationality
    collective 13, 24, 33, 151
    communicative 27, 152
reality 55-56
reference groups 121, 123, 125
release stage, panarchy 64
reorganisation stage, panarchy *64*, 65
resilience 62, 76, 155
risk communication 110
River Clyde, salmon 51
River Thames, waste disposal 3-4

salmon
    fish farming 9, 35, 39
    wild 34
        hazards of migration 35
        as indicator of water quality 34, 51
        parasites 35, 39
        protection 35
        River Clyde 51
salmon lice parasites 35, 39
saturation curve 59
scenarios
    importance of dialogue 116-117
    SPICOSA 21, 69-72, 77, 100, 155
    System Output 128
Scheveningen, Netherlands 4-5, 25
science
    hard view of 55, 67, 133
    need for interdisciplinarity 142

soft view of 55, 133
science and society
  communication 103-105
    change management 109
    cultural differences 105-107
    decision-making support tools 114
    deliberation support tools 114-117
    dialogue 107-110
    participation tools 112-113
    risk 110
    scientific information 109
    sustainability 108-110
Science-Policy Interface 22, 23, 105, 113, 138, 141, 142, 143-144
sciences, role in policy-making 42-45, 50
Scottish Enlightenment 79
Scottish Environment Protection Agency 16
sea level change 4
sea lice see salmon lice
sea urchins, and seaweed, trophic cascade 61, 76
sea-otter, control of sea urchins, trophic cascade 61, 76
self-regulation see homeostasis
sewage, and eutrophication 139-140
sewage leakage 32, 47, 48
sewage treatment, scenario 69-70, 71
Shanghai, coastal population 3
shellfish
  aquaculture
    Lagoon of Venice 38-39, 82-83
      conceptual model 84-89
      institutional map 49, 99
    prehistoric diet 3
    see also clam fisheries; mussel farming;
  oyster farming
simulation model 22, 89-95, 125, 139, 155
Smith, Adam
  on systems 79
  'the invisible hand' 12
social learning 110, 116, 128, 138, 141-142, 148, 155
Social-Ecological System (SES) 11, 13, 45
  dysfunction 137
  see also Issue
sociology 13
soft systems 55, 57, 65, 133

Soft Systems Methodology 27, 57, 155
  and General Systems Theory 138
South America, coastal zone vulnerability 145
SPICOSA (Science and Policy Integration for Coastal System Assessment) 20-24
  ExtendSim computer program 92
  scenarios 21, 69-72, 155
  study sites 21
    Basin de Marennes Oléron, oyster farming 129-130, 131, 132
    cultural differences 105-106
    Himmerfjärden, nitrogen runoff 47, 48, 100, **120**, 121, 133
    Lagoon of Venice, clam fisheries, modelling 39, 82, 85-89
    Mar Piccolo mussel farming 121-125
    Oder Estuary marshlands 127
Systems Approach Framework (SAF) 21-24, 22
  communication 105-106
  communications space 50
  hard and soft systems 57
  Loch Fyne 49
  Mar Piccolo mussel farming 121-125
stakeholder-issue mapping 50, 118, **119**, 134
stakeholders 17-18, 22-23, 27, 49-50
  communication to and with 103, 105-117, 127-128, 141
  engagement 18, 110-112, 141-142, 143, 148
  forums 114, 134
  and policy making 39-40, 49
state variables 66, 98, 102
STELLA computer program 92-95, 102
stoccafisso 33
Strandweg, Scheveningen 4, 5
successional climax 64
sustainability 8-9, 13, 26, 67
  communication of 108-109
  and Systems Approach Framework 142-143
System Appraisal 22, 126
System Design 21-22, 79, 84-89, 125-126
System Formulation 22, 125-126
System Output 22, 127-129
system states 62-63, 65-67

systems, management of 67-69
Systems Approach 53-75
    and communication 104-105
Systems Approach Framework (SAF) 21-24,
    22, 75, 79-101
    coastal zone social-ecological
        dysfunction 138-143
    communication and engagement theory
        105-117, 133
        decision support tools 114-117
        deliberation tools 113-114
        participation tools 112-113
    making an application 117-133
        analysis of problem 118
        further resources 147
        institutional map 118
        interpretative analysis 126-127
        Issue Identification 121-125
                eutrophication,
                Himmersfjärden 120, 121
                Mar Piccolo mussel farming
            121-125
        Output Step 127-129
        reference groups 121, 123, 125
        System Appraisal 126
        System Design 125
        System Formulation 125-126
        visualisation 129-130, 131, 132
    models 81
systems theory 18-20, 27, 53-75
    emergent properties 19, 152
    open 54
    properties 54
    sustainable coastal zone 20-24, 27-28

Taranto, Mar Piccolo mussel farming
    121-125
TechnoGardens scenario 72, 74
thermodynamics, systems states 65-66, 67
thermostat-heater systems, feedback loops
    19, 59-60
'think global, act local' 9
timesteps 89-90
tipping points 38, 63
Tokyo, coastal population 3
tools
    decision-support 114-117

deliberation 113-114
    participation 112-113
tourism, beach development 145
trade, cities 3
transaction costs 32, 156
transdisciplinarity 126, 127, 135, 142, 148
transformation costs 32, 156
Transformation process, CATWOE tool 119,
    134
transport, cities 3
trophic cascades 61, 76

United Nations 14-15
    Environment Programme 2
United States of America
    Harmful Algal Bloom and Hypoxia
        Research and Control Act 16-17
    Oceans Act 16-17
Users 45

variables 58, 66, 91, 98, 156
Venice
    Lagoon of 80-81
        acqua alta 81, 101
        clam fisheries 38-39, 82
            computer program 91
            conceptual model 82-89, 101
            institutional map 49, 99
            STELLA computer program 92-95
        socio-ecological system 53, 54, 57
        water-level models 81, 100
virtual systems 57-58, 71, 139, 156
visualisation 114, 125, 129-130, 131, 132

waste, flotsam and jetsam 6
waste disposal, cities 3-4, 25
Water and Environmental Services
    (Scotland) Act 16
water quality 32
whales, stakeholders 17
wind farms 6
worlds, work of Karl Popper 56, 143
Worldview 119, 134